Methods to Extend Mechanical Component Life

Lessons Learned with Space Vehicle and Rocket Engine Components

Dieter K. Huzel

Published and distributed by

AIAA

American Institute of Aeronautics and Astronautics
The Aerospace Center
370 L'Enfant Promenade, SW
Washington, DC 20024-2518

American Institute of Aeronautics and Astronautics, Inc.
370 L'Enfant Promenade, SW, Washington, DC 20024-2518

ISBN 1-56347-072-1

Preface

Since the ascent of modern, liquid propellant rocketry during World War II, one overriding thought governed the design and development of all launch vehicles and their components: to bring them reliably off the launch pad, on through powered flight to the programmed terminal engine cutoff. With the attainment of the power-off point of their main propulsion systems, these vehicles entered a ballistic, orbital, translunar or return (transEarth) trajectory. At this point, the propulsion systems and all of their components had served and expended their whole life function, except for, if applicable, the vehicle's low-thrust attitude control systems.

This "one shot" precept completely pervaded launch vehicle technology. An additional, most significant inference of this precept was the fact that the hardware involved, for all practical purposes, was irretrievable, either because it was destroyed during re-entry or impact, or because it remained beyond reach, by staying in orbit, or by coming to rest on a deep ocean floor. As a result, the designers and developers of the systems were left without real-life feedback, and they were often forced to rely on inadequate substitutes, such as telemetry, overdesign, unrepresentative ground tests, or gruelling qualification programs. As the launch systems of the past grew larger and more complex with corresponding increases in cost, congressional criticism, often based on narrow thinking, threatened to obscure the immense long-term values of the space program. In response to this situation, the development of reusable, recoverable space launch systems, which heretofore had mostly been treated in technical journals and in papers to symposia, became a matter of policy with the National Astronautics and Space Administration (NASA). As a first significant step, in-depth studies were initiated which eventually led to a defined concept for the Space Shuttle. Other needed systems can be expected to follow.

Of necessity, the first years of outlining, comparing, and evaluating these new systems were devoted and confined to systems and vehicle conceptual levels. As these studies progressed, often projecting very high reusability rates, those in government and industry who had been involved in earlier space hardware design, qualification, and procurement became increasingly concerned about the growing need for work at the component level. This concern was caused not only by the projected high mission numbers—a hundred flights (or more) without overhaul, instead of just one—but also by the lack of experience, by the lack of data, and simply by the lack of "how to go about it" plans. Equally alarming to them was the dissipation of skills, sometimes to the point of complete dissolution of highly specialized suppliers, and the irretrievable loss of experience and know-how at the conclusion or termination of major space programs.

This discussion introduces the reader to the field which is often referred to as "Long-Life Technology." It concentrates on practical, detailed recommendations for the design of reusable mechanical and electromechanical launch vehicle components. These components are sometimes referred to as acyclic (reciprocating) devices, as distinguished from cyclic (rotating) machinery, of which the turbopump systems of liquid propellant rocket engines are a prime example. The rotating machinery, too, plays an important role in the longevity of rocket engines—or for that matter in numerous other fields of technology—but it represents an area that merits special, separate treatment which would be beyond the scope of this book.

Special attention has been given to clarify every definition in the pages that follow. It is important for the reader to know what the author's interpretation and use of a term is, especially if conflicting definitions can be found in the existing literature. From time to time, the reader may wish to refer back to one definition or another.

The author would like to impress on the reader his belief that the design and development of reusable launch vehicle components do not require technological breakthroughs. In the pre-Saturn/Apollo and Shuttle days, anyone in charge of a liquid propellant rocket launch who would have tried to schedule a launch-time by day, hour, and minute, *6 months ahead in time*, would have caused much laughter in the space industry's engineering offices across the country, and serious questions would probably have been raised about his fundamental understanding of the realities of liquid propellant rocketry. Yet this type of schedule has been met many times during the Apollo/Saturn V launch program. Such scheduling is taking place again with reusable space launch systems, of which the Shuttle is currently the prime representative. Beyond the latter system, space stations, Moon bases, Mars missions, and various new rocket-powered space transportation systems are in the planning stages. Their components can and will be designed, developed, built, and ultimately flown, or used reliably. The realization of this goal will require meticulous attention to detail, unbiased application of all experience available, sensible test programs, dedication, and common sense.

Special recognition goes to Edward D. Storms of Rockwell's Space Division, who has made major contributions to Long-Life Technology and who advanced valuable suggestions for this publication.

The author is indebted to the manifold support he received from Rockwell International, Inc., and especially from its Rocketdyne Division.

Dieter K. Huzel
July 1993

About the Author

Dieter K. Huzel was born and raised in Essen, Germany. In 1937, he received his Masters Degree in Electrical Engineering from the Technical University of Stuttgart, Germany, after which he joined the Siemens-Schuckertwerke in Berlin. In 1942, he was drafted into the German Army and spent a year at the Russian fronts. In early 1943, he was transferred to the Army's Rocket Development Installation at Peenemuende. His position included that of manager of the V-2 static test firing and launch facility and, subsequently, of Technical Assistant to Dr. W. von Braun. After the war ended, he came to the United States under Government contract, as a member of Dr. von Braun's group. His experiences during this period are vividly described in his book "Peenemuende to Canaveral" (Prentice-Hall, New Jersey, 1962, and republished by Greenwood Publishing Group, Inc., Westport, Connecticut, 1981). In 1950, Mr. Huzel joined the Rocketdyne Division of North American Aviation, Inc., (now Rockwell International, Inc.), Los Angeles, where he held various supervisory positions in the development of large liquid rocket engine development projects. In 1960, he transferred to the Space Division of Rockwell International, where he was appointed the Technical Director, Propulsion Coordination, in the Apollo Program's Saturn S-II stage project, and to other subsequent supervisory positions. During these years, he coauthored, with D. H. Huang, the textbook *Design of Liquid Propellant Rocket Engines* (NASA SP 125, 2nd Ed, 1971). After 26 years with North American Aviation/Rockwell International, Mr. Huzel retired on June 30, 1976. Mr. Huzel is an Associate Fellow of the American Institute of Aeronautics and Astronautics and the coauthor (with D. H. Huang) of the AIAA Progress in Astronautics and Aeronautics Volume 147, *Modern Engineering for Design of Liquid-Propellant Rocket Engines* (1992).

Table of Contents

Chapter 1. "Long Life" Defined **1**

Chapter 2. Causes of Diminished Component Life **3**
Principal Component Degradation Mechanisms 3
The Nature of Wear 5
Materials Aging 7

Chapter 3. Means to Extend Component Life **9**
Improved Component Wear Analyses 10
The Interface Principle 10
Wear-Analysis Methodology 11
Specimen Selection for Endurance Test 11
Data Books 11
Analysis Team Member Selection 12
Endurance Tests 12
Component Disassembly 13
Analysis 13
Wear Rating 14
Wear-Rating Procedure 15
Materials Aging 25
Age Life of Common Materials 25
A "Family Tree" of Nonmetallic Materials 36
Accelerated Aging 38
Thermal Analyses 39
Prospects of Accelerated Aging 39

Chapter 4. Some Words of Caution **43**
Weight and Size 43
Sophistication 43
Material Selection 44
Materials Processing and Fabrication 44
Component Timing 44
Maintainability 44
Ease of Checkout 44
Accessibility 44
Mismatch 45

Communications 45
Late Changes 45
Mishandling 46
Safety 46
Qualification 46
Corrosion 46
Stress Corrosion 46
Fracture Mechanics 46
Contamination 47
Creep and Drift 47
Flow Dynamics 47
Impact Loads from Waterhammer 47
Environment 48

Chapter 5. Interface Design Concept Tabulation 49
1. Reciprocating Shaft Seals 54
2. Reciprocating Piston or Poppet Seals 55
3. Rotating Shaft Seals 57
4. Rotating Main Closure Seals 57
5., 6. Reciprocating Poppet, Ball, or Piston Seats 58
7., 8. Reciprocating Guides 60
9. Rotating Shaft Journal Bearings 61
10. Rotating Shaft Ball and Roller Bearings and Ball Screws 62
11. Rotating Pin Joint Bearings 62
12., 13., 14., 15. Thrust Bearings, Miscellaneous Guides/Bearings, Stops, and Ball Locks 63

Chapter 6. Warnings and Predictions of Component Failure 65
Signature Analysis Methodologies 67
Observations and Conclusions 69

References 73

Appendix A: Interface Design Concept Tabulation 75

List of Figures

2-1. Mechanisms Contributing to Component Degradation (Wear-Out) 4
3-1. Wear Analysis Example No. 1: 1 1/2-in. Helium Pressure Regulator 16
3-2. Example No. 1: Interface Wear Rating 17
3-3. Example No. 1: Estimated Extended-Life Potential 17
3-4. Example No. 1: Estimated Prolonged-Life Potential 17
3-5. Example No. 1: Estimated Long-Life Potential 17
3-6. Wear Analysis Example No. 2: 4-in. J-2 Engine Main Fuel Valve 18
3-7. Wear Analysis Example No. 2: 4-in. J-2 Engine Main Fuel Valve, Interface Details 19
3-8. Example No. 2: Interface Wear Rating 20
3-9. Example No. 2: Estimated Extended-Life Potential 20
3-10. Example No. 2: Estimated Prolonged-Life Potential 20
3-11. Example No. 2: Estimated Long-Life Potential 20
3-12. Wear Analysis Example No. 3: 3/8-in. Solenoid Valve 21
3-13. Example No. 3: Interface Wear Rating 22
3-14. Example No. 3: Estimated Long-Life Potential 22
3-15. Polymeric Materials—Terminologies and Relationships 37
5-1. Typical Interface Design Concept Tabulation Format 51
5-2. Typical Modified Versus Unmodified Interface Configurations 53
6-1. Solenoid Valve Oscillogram—Pickup Cycle 66
6-2. Tracing of Solenoid Valve Pickup Voltage, Current, and Accelerometer Oscillograph Records 67

List of Tables

3-1. Recommended Modifications for Increasing the Operational Life of the 1 1/2-in. Helium Pressure Regulator 23
3-2. Recommended Modifications for Increasing the Operational Life of the 4-in. J-2 Engine Main Fuel Valve 24
3-3. Recommended Modifications for Increasing the Operational Life of the 3/8-in. Solenoid Valve Assembly 24
3-4. Room Temperature (Earth Environment) Storage Life of Common Materials Used in Spacecraft and Launch Vehicles 26

Chapter 1

"Long Life" Defined

A VALID question would be: How long is long? A general answer is: The length of life of a multimission rocket engine component is that which is delineated by its specification. In turn, a component's specification is closely linked to the mission profile of the system of which it is to become a part. A specification usually spells out numerous requirements. Most important among these requirements for trouble-free, Long-Life capability and its demonstration are the stipulations for the total number of "in-spec" operational, cumulative "on" times (e.g., a turbopump) and/or the number of cycles and maximum permissible leak rates (e.g., a valve). It is the latter category with which we are primarily concerned here.

Those preparing the specification must realistically consider the applicable vehicle mission profile, and at the same time they must avoid overly stringent, unrealistic or "panicky" requirements which would merely add to cost, without any true benefits. These considerations also apply to the specifications for the component's qualification tests (proof of soundness of design) and acceptance tests (soundness of individual production units). Some thoughts about what these tests cannot do are discussed further in Chapter 3. In that chapter we consider methodologies of assessing the life expectations and capabilities of engine components and specifically, their dynamic interfaces. These methodologies have important economic and cost-savings implications, because there may be instances where an existing component can be uprated cost effectively to new, above-specification applications, as a result of tests and wear analyses.

In order to outline the spectrum of "Long Life" and to explain the terms used in our discourses, the following definitions are set forth:

Extended Life — This term shall mean the capability of a component to perform a limited multiple of the number of missions for which it was originally designed, without out-of-specification performance degradation. For the purposes of this discussion, assuming we are dealing with hardware originally designed for single missions, the term *multiple* shall mean several—typically three—complete equivalent operational missions without refurbishments or replacements. It also implies that requalification would not be required, based on the following premise:

Recommended minor modifications have been applied and verified.

The original qualification entailed multiples of single-mission operational life greater than three.

Routine inspection of flight records and/or hardware is possible and performed between missions.

It is further implied, although it is not a condition, that the component life may be further extended if performance during the basic three missions so warrants, and if it is acceptable to all concerned.

Prolonged Life — This term shall mean the capability of a component to perform a specified number of missions significantly greater than that for which it was originally designed (but less than *Long Life*, the next item below), without out-of-specification performance degradation. The term *Prolonged Life* is introduced to permit possible reconsideration of components that initially were not recommended for Long-Life applications but that subsequently demonstrated capabilities greater than Extended Life (though less than Long Life). In some cases, life extension may be accomplished through maintenance procedures, adjustments, or replacements (scheduled or unscheduled). Compared to a completely new design, this approach offers reduced cost.

It is expected that minor-to-nominal design modifications and to-be-defined demonstration (qualification) testing will be required before operational use of an existing component for Prolonged Life applications.

Long Life — This term denotes the capability of a system, subsystem, component, or part (mechanical and electrical), or the materials or the structural members of which they are composed, to function reliably within specifications, before and after installation, accurately and without failure, with minimum or no service, continuously or intermittently, for a specified period of years or number of missions, in any environment to which launch vehicles, spacecraft, and extraterrestrial installations, manned or unmanned, may be exposed.

It may be helpful to define further several of the terms used in the foregoing definitions:

Performance Degradation — This term applies to the gradual or erratic, time-dependent drift of specified performance parameters of a component (1) away from acceptance-test values, or (2) outside of tolerance values specified for the duration of its operational life. Typical parameters are response times, travel time, response pressure, and leak rates (internal and external). Situation (1), while not detrimental as long as it is within tolerances, may be indicative of impending situation (2).

Minor Modification — This term is confined to limited rework of existing detail parts of components. Rework may include minor dimensional changes, material changes, surface-finish changes, and changes in surface treatment (e.g., platings, dry film lubricants) as required to remedy adverse wear conditions.

Nominal Modification — This term denotes more extensive rework of existing detail parts or the addition of new detail parts within the envelope confines of an existing component but excludes redesign through the application of New Concepts (see "New Concept," Chapter 3).

It is conceivable that in some projects, terminologies other than those enumerated above are preferred or are in use. Terminology will not be a problem as long as all participants agree on the meaning of each term chosen and use it consistently.

Chapter 2

Causes of Diminished Component Life

IT is common knowledge that nothing lasts forever. But it is also a fact that some things last longer than others. This truth is well known to consumers who are comparing generic devices, such as refrigerators, vacuum cleaners or automobiles, *and at the same time are considering their cost.*

Although the consumer may find assistance with his cost-effective selections in consumer-oriented publications, the rocket engine component designer has no such recourse. It may occasionally be possible to include a modified and/or uprated "off-the-shelf" component in a new engine design. However, most engine elements have to be newly specified, designed, and manufactured for a particular configuration.

Sometimes, when a commercial device or a rocket engine component fails, the tendency is to dismiss the problem as a "random failure." That reaction, however, is a subterfuge. The engine designer and developer should be well advised that *there always is a cause, which must be identified and eliminated.* Accordingly, in this section we shall discuss the potential causes of diminished component life; then in Chapter 3, we shall describe methods to identify these causes, to extend life, and to prevent recurrence of failures.

The operational life of a rocket engine depends on the performance of its components, and it may be just one component in the assembly that creates the eventual engine problem. In fact, it may be just one feature of that one component that constitutes the "weak link" leading to failure. These failures predominantly occur at the dynamic interfaces between moving parts; to a much lesser degree they occur in other locations. Methods of identification of the weak links will be the principal subject of our discussions.

Principal Component Degradation Mechanisms

The failure of a component to achieve specific operational life may simply be the result of a deficient design, which may be caused by lack of experience, poor choice of detail design concepts, materials, tolerances, and surface finishes, unclear manufacturing instructions, and the like. A poor design may also be traced to ignorance about, or a disregard of, the component's operational environment. (However, this probably does not happen often.) Conversely, the most competent designer cannot guarantee that his component will function flawlessly from the outset. He needs development tests and analyses to assure himself and others that his device has no weak links to cause premature component degradation—and to ensure that in the event any flaws were detected, they would be eliminated.

Component degradation (sometimes referred to as *component aging*) is caused by two principal mechanisms:

Wear, as a function of component usage or operational "cycles"

Materials Aging, as a function of exposure to various environments and of elapsed time.

Both are used here as generic terms and comprise several subeffects, no doubt mutually influencing one another to some extent. Figure 2-1 illustrates these principal interrelationships.

In addition to the mechanisms shown in Figure 2-1, the designer, the manufacturer, and the user must pay attention to specific additional causes that may affect component life. Among these causes are the following:

Workmanship (skills, pride, quality control, inspection)
Corrosion (moisture, materials selection, cleanliness, protection)
Stress corrosion (understanding, materials selection)
Electrolysis (dissimilar metals, moisture)
Contamination (clean rooms, protection)
Hydrogen embrittlement (fracture mechanics, materials selection)
Vibration (handling, installation, environment)
Handling (competence, training, equipment)
Packaging (expertise, protection)
Storage (environment, protection, security)
Radiation (materials selection, protection)

During a technical discussion, Wernher von Braun once observed: "If a technical meeting is

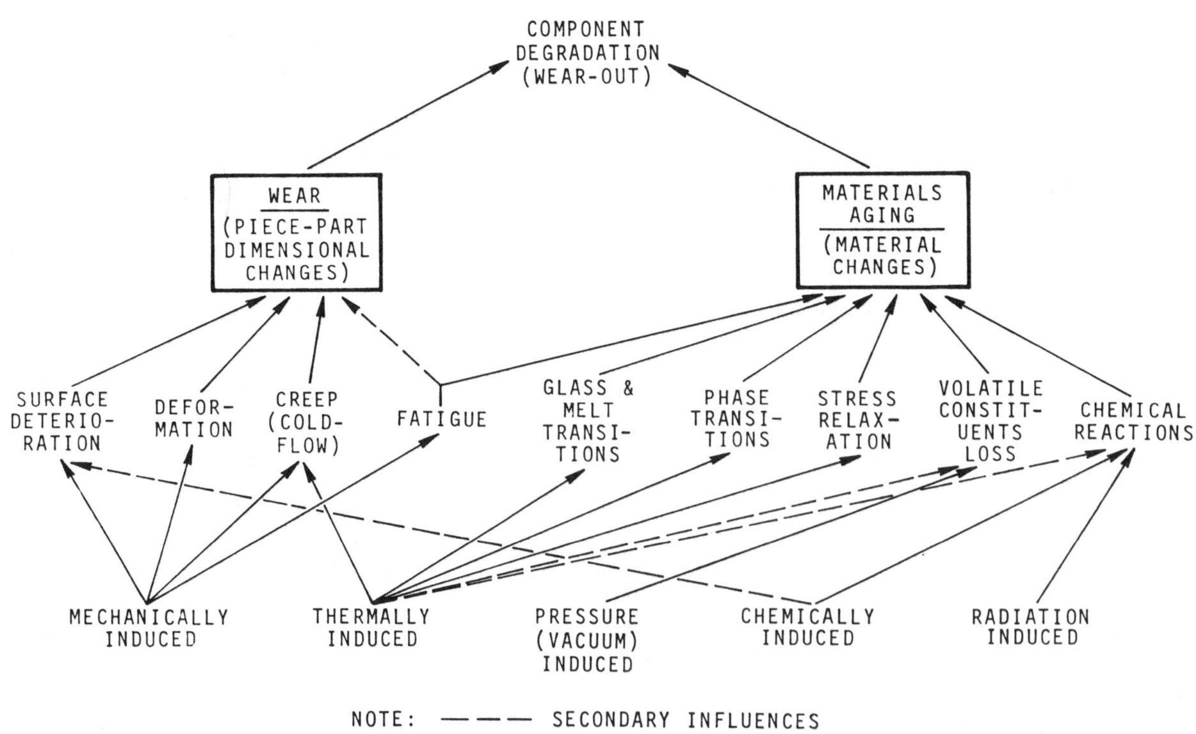

Figure 2-1. Mechanisms Contributing to Component Degradation (Wear-Out)

becoming sluggish and aimless, all you have to say is: 'Now, what about the vibration?' There will be a lively response. Just in case another slowdown looms, just ask: 'And what about contamination?'" While this was a tongue-in-cheek remark, it was not off-target. Both agents—vibration and contamination—are powerful contributors to premature component failure. Conscious design and proper mounting (shockmounting) of components are the principal means of protection against vibration effects.

For the prevention of contamination effects, we must consider external sources and internal sources. The magnitude of surface deterioration is often directly related to the type, size, and density of the contaminants. Self-generated contamination is considered by many as the most prevalent form encountered in components. Under the heading of self-generated contaminants fall such items as parent material burr break-off, flaking of platings, parent material fretting under vibration, parent material spalling under impact loading, and disintegration of nonmetallics. The identification and control of self-generated contaminants are primarily within the area of responsibility of the component designer. System-introduced contaminants, on the other hand, such as moisture (icing) in cryogenic components, are often more difficult to combat because of the inability of the analyst to separate the source—externally introduced or self-generated—because the type and source of the original contaminants are often masked by the products of subsequent wear. From this standpoint, the analysis of wear and contamination may be inseparable.

In cases where gross foreign particle contamination is established, no further consideration should be given at the component level, on the premise that the avoidance must be achieved at the system level, without component redesign.

The two principal causes of component degradation are discussed in more detail as follows.

The Nature of Wear

Wear is the one condition considered to be the most prevalent in influencing performance longevity. It is a major parameter in the analysis of the processes of component deterioration (component aging). Wear affects both metallic and nonmetallic materials. Continuing efforts to better understand, analyze, and reduce the processes of wear, therefore, will materially contribute to the achievement of the ultimate goal of designing, developing, and producing long-life components with confidence. We want to emphasize that all components must be expected to exhibit, during their design lifetime, a certain amount of normal tolerable wear in the often severe, predicted operational environments of projected future space and launch vehicles. It is the abnormal, unpredictable, and excessive wear that is of concern here. The aim, therefore, must be the avoidance, or at least a substantial reduction, of abnormal wear, to assure component performance life well beyond a safe, minimum limit.

As used in this discussion, wear is defined as any mechanical abrasion, erosion, deformation, or other deterioration of surfaces in contact under relative motion. The effects of space environments and of certain forms of erosion, such as those caused by chemical and electrochemical reactions, and by electrical discharge material displacement, are considered only to the extent that they may contribute to subsequent mechanical surface deterioration.

In many factories, materials laboratories, and engineering groups that are concerned with component failure analyses, design evaluations, and hardware maintenance, a number of terms are being used to describe the various symptoms of wear, especially for mechanically induced surface deteriorations. Because these terms are not being used uniformly in daily activities and in literature, the following definitions summarize the principal interpretations that have been used herein:

Burnishing — Burnishing is defined as the rubbing or polishing of contacting surfaces under relative motion. This process results in lustrous or glossy areas of wear with extremely fine particle generation. In the case of plastic materials, burnished surfaces may exhibit areas of discoloration and deposits of loose or adhering parent material in what appears to be a smear.

Scuffing — Scuffing is usually associated with plastics and other relatively soft materials. It is defined as localized, randomly oriented tearing or shredding of a surface, analogous to some extent with the galling of metals.

Fretting— Fretting is defined as a gradual erosion of contacting surfaces under small oscillatory motion. Since fretting has been observed to occur over a wide range of loads, its magnitude appears to be more a function of cycles than of the imposed loads. For example, surface fretting is a common condition encountered under vibration. Small-size particle generation usually accompanies this form of surface erosion.

Spalling — Spalling is defined as a rapid erosion of contacting surfaces as a result of exposure to high-impact cyclic loading. Brittle materials are considered more susceptible to this type of surface deterioration. Large-size particle generation (flaking) may accompany this type of surface erosion.

Galling — Galling is defined as extreme deterioration of mating surfaces as a result of high-sliding unit loading, introduction of contaminants, or lack of proper lubrication. This term also applies to the intermingling of mating-surface materials to the ultimate extent of cold welding or complete seizure. From this standpoint, galling is more closely aligned with adhesive wear than with abrasive wear (see the definitions of adhesive wear and abrasive wear). Any of the lesser surface deterioration conditions may eventually be manifested in complete seizure or galling of mating surfaces.

Brinnelling — Brinnelling is defined as plastic surface deformation which results from loading beyond the elastic properties of materials. This condition often accompanies misalignment or high-impact loading.

Crazing — Crazing is defined as surface deterioration in the form of minute surface or subsurface cracks, possibly incurred as a result of exposure to rapid temperature changes or to rapid load reversals.

Adhesive Wear — Adhesive wear is defined as an attraction between surface atoms of two contacting materials, leading to transfer of material from one surface to the other and eventually to the formation of loose fragments. Presently, two major theories for adhesive wear are widely accepted. One is known as the *welding hypothesis*, and the other is known as the *theory of metal transfer*.

Abrasive Wear — Abrasive wear is defined as the removal of material by the filing action of a hard, rough surface against a softer one, or by abrasive grains trapped between smooth surfaces. The great bulk of wear upon lubricated metal surfaces is caused by entrapped grit being pressed against the metal. Abrasive grit is found almost everywhere in the form of ordinary dust. This dust or grit, when pressed against the surface of a metal, plows a furrow in the surface, forces metal into the space between the surfaces, and thus decreases the clearance. Rough surfaces, grinding cracks, or scratches from grit are stress-raisers that lead to deterioration by fatigue.

Materials Aging

The selection of suitable materials is one of the most important aspects of successful long-life component design and development. However, particularly for *nonmetallics*, the task has frequently proven difficult because of the absence of reliable materials data for the long operational periods envisioned for space vehicles and systems of the future (10 to 15 yr or more). A general assumption is that it is predominantly the nonmetallic materials which, apart from mechanical effects, such as wear, deformation, and creep caused by component assembly and operation, are subject to an aging process induced by elapsed time and environment. By contrast, the materials aging of *metals* from these causes, as distinguished, for instance, from corrosion, is usually not considered by the designer to be a problem.

Moreover, as long as the aging effects of nonmetallic materials do not interfere with the component's performance during its design lifetime, they should be of no concern, either. To ensure this, various means of predicting material integrity are being pursued and have been reported in publications. One of these means is the performance of *accelerated life tests*; the nature and limitations of these tests will be discussed in Chapter 3. Another means is the study of the behavior of materials through real-time (and thus time-consuming) tests, either by analyzing materials that had been stored for many years under known conditions or by planning and performing controlled tests.

In Chapter 3 we discuss experience gained and research and tests performed with various, mostly nonmetallic materials. These data should help to prevent the inadvertent choice of a material unsuitable for a given application.

Means to Extend Component Life

FOR application in single-launch, "expendable" vehicles, most components were generated within the rigid confines of conventional design, development, and verification test (qualification and certification) phases, without due consideration of their ultimate performance capability (endurance limit). And too frequently, most of the stringent schedules imposed on component design and development programs led to premature component production. At that point, the component became a closely controlled entity (end item). Any design improvements subsequently proposed were subjected to severe scrutiny to avoid production impact. Generally, only mandatory changes were processed. Detected design problems of lesser immediate concern were often postponed until their resolution became mandatory and often dictated a major retrofit as well, because a greater number of components had accumulated in the meantime. In fact, component retrofit was occasionally accomplished on a vehicle that was literally on the launching pad. Often, qualification test programs were conducted, concurrent with initial component production. These test programs usually experienced sporadic interruption—they were stopped and restarted anew—as a result of production changes to the basic component configuration. Conversely, component production was adversely impacted many times by major design changes which had to be implemented to resolve component performance deficiencies detected during the verification testing of earlier changes. This entire vicious cycle culminated in *an inefficient utilization of available time and funding*.

The qualification test programs, themselves, contributed further to the overall complexity surrounding single-launch component development because these programs were essentially "brute force" demonstrations of a component's capability to meet more or less arbitrarily scaled-up multiples of anticipated operational life. Again and again, the relationship of qualification test exposure and the required specimen life to actual operational life requirements was highly questionable. The stereotyped structuring of these test programs included costly, time-consuming tests, usually unrepresentative of anticipated operational usage, such as shock, temperature-humidity, dust and fungus, and certain thermal profiles. Most significantly, these test programs concentrated on qualification of the component assembly, commonly represented by several specimens, without thorough consideration of subassembly or dynamic interface performance. In turn, performance anomalies were held against the complete component assembly, rather than against the particular dynamic interface(s) at fault or, to a lesser degree, against the selection of an unsuitable material. Consequently, valuable interface concepts were discarded when a component was abandoned as a unit, even though, quite regularly, no more than one interface problem was prohibiting its qualification.

In many instances, periodic design reviews and other reliability-related activities have demonstrated only marginal effectiveness in isolating potential component problems. Most of these assembly-oriented reviews are cursory in nature and lack sufficient depth to uncover subtle interface-related problems. Other statistical reliability methods provide component longevity numbers of little utility because they were developed without the benefit of knowing a component's ultimate life capability. Only the failure analyses of previously failed components, such as those which occurred during qualification tests, came closer to our proposed evaluation approach—analysis at the interface level. However, these analyses were limited in scope because the focus of each one was set on the malfunctioning interface (or small group of interfaces) and on the expeditious correction needed—without a look at all the other interfaces.

Improved Component Wear Analyses

The following passages outline an effective approach to component development for future rocket engine system applications through utilization of the Interface Principle (weak-link identification), encompassing endurance tests followed by methodical wear analyses, design improvement recommendations, redesign, and proof tests. This approach shifts the emphasis from the assembly level to the subassembly or interface level, for all dynamic interfaces. Note that while the approach will generate the greatest benefits during design and development of a new component, before the start of production, it will also be quite useful if applied to production components which failed during qualification, acceptance testing, or operational use. *Moreover, this methodology has significance, well beyond the realm of liquid propellant rocket engines.*

The Interface Principle

The Interface Principle, as defined here, subscribes to the Weak-Link Concept, i.e., the simple premise that the operational longevity of a mechanical or electromechanical component, as a whole, is directly related to the longevity of its weakest operational interface(s) between moving parts. Accordingly, the study of a component at the operational or dynamic interface level is the basic working mechanism in the application of this principle. An operational or dynamic interface is defined as the most elementary combination of mechanical elements in a component, providing surfaces in contact and under relative motion. *Unlike mere failure analysis (which tests only those features associated with performance anomalies), this concept provides also for the evaluation of a component's good features.*

The Interface Principle addresses itself to component operational life as a function of wear-out (see Figure 2-1), rather than as a function of structural failure. Structural failure is addressed briefly in a few places in this discussion, but it is not considered a major problem for long-life design because it can usually be identified, diagnosed, and corrected in a straightforward manner.

Component wear-out is defined as either the mechanical binding or seizure of an interface, or as "out-of-specification" deterioration of interface performance (i.e., sluggish operation, excessive leakage across a seal interface, etc.). It is assumed that the analytical tools available to the component designer and to the materials, stress, and dynamics analysts have already been utilized to identify and remedy any piece-part structural weaknesses within the component, in accordance with standard practices. Exclusive of piece-part structural integrity considerations, the operational life of a component is considered to be directly related to wear-out of its weakest mechanical interface. It is recognized that the sometimes rapid structural breakdown of interfacing surfaces may be the result of Materials Aging (changes from the application of loads, environmental exposure, elapsed time, or a combination thereof). For example, polymers may exhibit drastic changes in mechanical properties because of changes in crystallinity resulting from loading and/or exposure to high temperature environments. (This change, along with other related aspects, is addressed further in connection with Materials Aging.)

Component life extension, then, is assumed to be a function of early identification and resolution of its weakest mechanical interface. In turn, further life extension is deemed possible upon subsequent identification and resolution of its second and third weakest mechanical interface, and so on.

Wear-Analysis Methodology

The road to the assurance that all weak links in a component have been identified and corrected, and that the component will meet all specifications for operational life, entails certain steps which are identified in the following discussion.

Specimen Selection for Endurance Test

As we pointed out, customary qualification test programs have serious shortcomings. Even if the components selected for these tests pass, it is never known whether they would have failed a few seconds—or a few cycles—later. Because a component which passes a qualification test is usually not disassembled, no information is generated that could even theoretically project how close the component may have been to failure.

The recommended alternative is the endurance testing of one specimen (but ideally of several specimens) of a given design, for long time-periods, and/or for large multiples of specified cycle life—preferably to ultimate failure—followed by controlled disassembly and detail analysis. The specimen may be the production prototype of a new design or an existing component, considered potentially capable of performing within specification in an application lasting longer than the original design life, perhaps, after minor, economical modifications have been made. In the case of an existing component, cost may also be reduced if specimens are available that have already undergone extensively documented operational use or have failed during that use, such as qualification test samples.

In some instances it may be desirable to evaluate the life capabilities of just one interface, or of several specific dynamic interfaces, or modifications to them. In this case an available, suitable component may be selected to serve as a test bed for the concept(s) under evaluation.

Data Books

Upon selection of the component specimen(s) to be wear-analyzed and before the next steps in the analysis are taken, a data book should be assembled. In the first section, design data and the operational history (if any) of the component to be evaluated should be recorded. Typical sections must include these topics:

A detailed description of the component
The principles of its operation

Supporting sketches and diagrams
Schematics (of the system in which employed)
Design and performance data
List of applicable documents (drawings, specifications, reports)
Failure history (if any)
Sketches identifying and numbering all interfaces
Interface detail descriptions
Predisassembly assessment of component type
Operational history of the selected specimen (if any)

As the analysis progresses, information on test procedures and equipment, test results, wear ratings, assessments, and recommendations will be added to the book. The early preparation of the listed data book sections serves two purposes: it provides information to prevent disassembly of a specimen with operational exposure insufficient for a meaningful wear analysis, and it serves to thoroughly familiarize wear-analysis team members with the component and its history.

Analysis Team Member Selection

Selection of team members (assessors) is the next important step. The members of each component analysis team should be selected from various disciplines and backgrounds. The teams may typically be composed of three to four members—a manageable number of participants consistent with the maintenance of good communications and handling of individual inputs. Although this requirement is not necessarily essential, team members should be selected on the basis of a component-hardware-oriented background. The team members may include representatives from Design, Test, and other laboratory functions, as well as Reliability and Engineering management. For purchased components, cognizant supplier personnel should be invited to participate in the wear analysis, when such participation would be beneficial in enhancing the component evaluation effort.

Apart from the evaluation team members, a skilled mechanic must be available to perform the post-test disassembly. If possible, this disassembly should take place in the presence of the team members, permitting early, overall observations of contamination, particle deposits, prior assembly deficiencies, and leaked hydraulic fluids. This disassembly may help with the assessment of subsequent observations at the interface level; it also will largely prevent the attribution of irregularities to wear, when, in fact, the irregularities might actually have occurred during the disassembly process.

Endurance Tests

We have previously emphasized the value of endurance tests which include tests to failure, in order to positively identify a component's weakest link interface, apart from the thorough wear analyses of *all* its interfaces.

The setup of the endurance tests for a given component depends on the component's function. Above all, the test must closely simulate the component's operational environment, where temperature and vibration are of primary importance. When specifications for the test setup are being arranged, the methodologies used for the component's qualification tests would certainly provide a valuable and cost-saving guide. But if they are not available, the methodologies used for comparable earlier components should be consulted. Generally, if supplied with complete component data as they are compiled in the Data Book, an experienced test laboratory will be able to prepare and conduct meaningful tests. These will entail cycling tests, usually a combination of operational (wet) and checkout (dry) cycles, and vibration tests, wholly or partially at cryogenic temperatures.

To the greatest extent possible and economical, the team members should familiarize themselves with the test setup and the equipment used, and from time-to-time, observe the tests as they progress.

Component Disassembly

The preparations for post-test disassembly involve these steps:

Selecting a "foreman" for the team, preferably the person most closely associated with the component;

Choosing a suitable disassembly location, i.e., a room where space can be set aside for a sufficient length of time, and team members can gather around a table for examination of the disassembled parts;

Drafting a disassembly sequence, later to become part of the Data Book;

Removing all external lockwiring (if any) of threaded fasteners;

Ensuring that a mechanic familiar with the component will be present, as well as the needed tools (including those specifically made for the component).

During disassembly, breakaway torques of all fasteners should be recorded. Similarly, all pertinent observations at this point must be listed and made part of the Data Book. Display and arrangement of the piece parts on a table will facilitate the analysis and photography sessions which follow.

Analysis

After a component specimen has been disassembled, all of its dynamic interfaces are closely inspected and examined by the team members in a preselected sequence. Each member notes and records his observations of wear, material deposits, and other abnormalities. Completeness of the records can be assured if a simple form is prepared and distributed for that purpose. *It is important that "little or no wear" is also recorded for a given interface.*

At the end of each interface examination, to avoid oversights, but not to otherwise influence or modify individual assessments, the team members should briefly discuss their observations. Since only magnifying glasses may be used during this examination and analysis session, the discussion may also lead to the decision to submit some of the interface mating parts to additional metallurgical or chemical laboratory analyses. Also, at this point, photographs, including microphotographs, should be taken.

Wear Rating

Following the receipt of all special analysis results, the team members are asked to combine these with their own wear ratings for each interface and to submit their findings numerically, on a scale from 1 to 10, to a coordinator. Details of this procedure follow.

In the early 1970s, Rockwell International's Space Division, under contract to NASA, performed, with the participation of several component suppliers, an in-depth study during which the Long-Life methodologies were applied with excellent results. This study,[1] which extended over several years, included these efforts:

Wear analysis of 10 different engine and propellant feed system qualification test specimens from the Saturn V single-launch booster vehicle program. The analysis was to answer this question: Although these components had passed qualification tests, would they be suitable for a (hypothetical) long-life application?

It was concluded that all specimens, following minor or nominal recommended modifications (see definitions listed under Wear-Rating Procedure), would be suitable for Extended Life, while six were also judged good candidates for Long Life. Our experience with these analyses showed that it would be beneficial to introduce an intermediate estimation, Prolonged Life (as previously defined and as applied in later analyses).

Selection of three of these components for recommended modifications of several of their dynamic interfaces, followed by endurance tests at cryogenic temperatures, then wear analyses, and finally an assessment of the effectiveness of the improvements. Two of these three components were subjected to extensive vibration testing at LN2 temperatures.

With one exception, all interface modifications resulted in a definite—and sometimes significant—increase in postverification test wear-analysis ratings. And as a result, the confirming endurance tests showed that an average useful operational life extension of one order of magnitude, compared to results obtained in earlier qualification tests, could be projected.

Wear analysis of six components, including two of the endurance test specimens. Of these, all six were found to be suitable for Extended Life and Prolonged Life—and four needed no further modifications. Five specimens were judged suitable for Long Life, providing that recommended changes were applied. Of special interest may be the specimen not judged suitable for Long Life. In this component, all 20 interfaces of a J-2 engine main fuel valve, which had participated in 76 engine firings and had been subjected to numerous "dry" actuations for a total of 1,308 cycles, were judged capable of Long Life, with a few minor changes, except one: the 4-in. lipseal, which required a new concept.

In another component of this group after wear analysis, a 2-in. butterfly valve, the lipseal (also a former qualification test sample), was again judged the weak link. Following refurbishment and the application of minor and nominal changes, the valve was subjected to endurance tests to prove at least Extended Life and Prolonged Life potential. However, the tests proved the value to have Long-Life potential as well!

Wear-Rating Procedure

To describe the wear-rating procedures successfully employed in the noted study (see Reference 1), we shall first define several of the expressions used during assessment and rating:

Minor Modification— This term denotes a change in surface finish, the change of a tolerance, or the like.

Nominal Modification —This term refers to a change of the material, or shape, of a piece part, or to similar modifications, not affecting other parts.

New Concept—This term, New Concept Recommended, conveys the judgment of the evaluation team that one or more troublesome critical interfaces will require major redesign or substitution of a new concept (e.g., a rack and pinion drive substituted for a crank). In this case, the design of a new component "from scratch" may often be more economical overall than a redesign of the investigated component, with an outcome that is uncertain.

Normal Wear—This term is defined as the presence of burnishing (polishing) and minor surface deterioration to the extent that allowable seal-leakage levels are not exceeded and that bearing and other frictional loads do not interfere with proper component operation. Because allowable leakage and performance limits may vary widely for similar components undergoing wear trade-off studies, the "normal wear" definition may have to be revised to reflect arbitrarily established allowable leakage and performance limits mutually representative of the components involved in the analysis. We anticipate that these limits may eventually be established to a common baseline, such as size or contact area.

Abnormal Wear—This term is defined as surface deterioration causing out-of-specification performance to the extent that complete mechanical malfunction or loss of sealing capability occurs. Again, out-of-specification performance may be eventually redefined as performance outside of arbitrarily established limits. Abnormal wear shall manifest itself in the form of excessive degrees of fretting or galling (and so on) of sealing, bearing, and other mating-contact surfaces. Each of these noted conditions may be further aggravated by self-generating and/or externally introduced contamination.

Each of the two main wear classifications listed above (normal wear and abnormal wear) should be subdivided into the appropriate degrees of wear described by common surface-deterioration terminology, such as burnishing, scuffing, fretting, spalling, galling, brinelling, and crazing. (These terms are defined in Chapter 2.)

The detail steps of the Wear Rating Procedure may best be described by using three representative examples (Figures 3-1, 3-6/7, and 3-12). All three specimens were actually analyzed during the Long-Life Study at Rockwell International's Space Division. The results of the analyses are presented in several supporting figures and three tables.

Control No.: S-IC/207-5, Saturn S-IC 1½ Inch Helium Pressure Regulator, P/N 20M32015 (Parker Hannifin P/N 5640061); S/N 06553G017802, (Component Data Book Supplement Number 1)

Test Exposure:	Preflight Certification Test
Test Cycles:	500 Cycles (200 @ Amb., 150 @ +165 F, and 150 @ -200 F)
Cycle Requirements:	500 Cycles (200 @ Amb., 150 @ +150 F, and 150 @ -200 F)
Vibration Exposure:	69 Minutes

NOTES: (For Interface Location Drawings below)
1) The numbers in the circles are Interface Numbers
2) Interface 23 was initially selected for evaluation; however, it was subsequently deleted since it is not a dynamic interface.

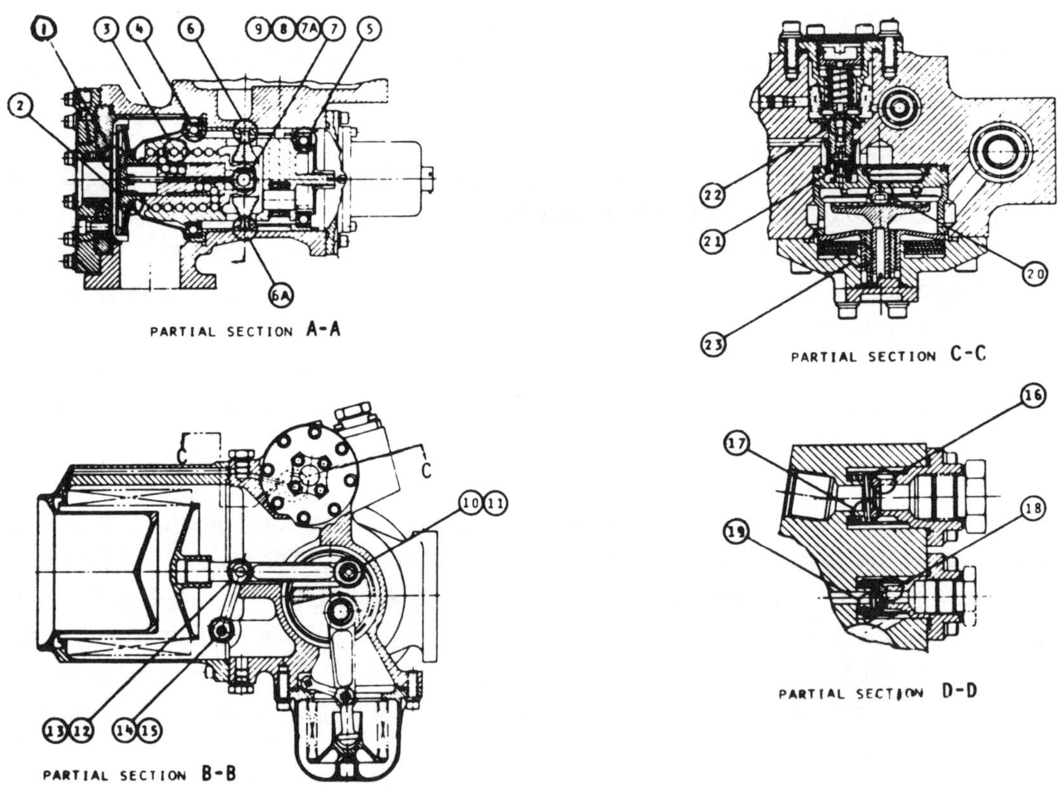

Figure 3-1. Wear Analysis Example No. 1: 1 1/2-in. Helium Pressure Regulator

In addition to providing sectional views, the figures list supporting information of significance to the analysts, including prior test exposure and design cycle-life requirements.

The first component (Figure 3-1) is a helium pressure regulator used for the pressurization of the Saturn S-IC (first stage) fuel tank. The second component (Figures 3-6 and 3-7) shows the main

CONTROL NO.: S-IC/207-5
SATURN S-IC 1½ INCH HELIUM PRESSURE REGULATOR
P/N 20M32015 (PARKER HANNIFIN P/N 5640061),
S/N 06553G017802

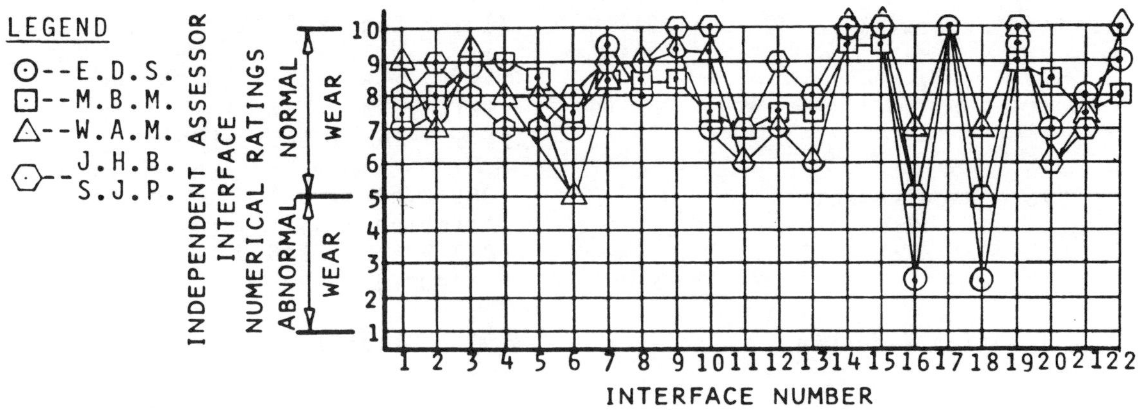

Figure 3-2. Example No. 1: Interface Wear Rating

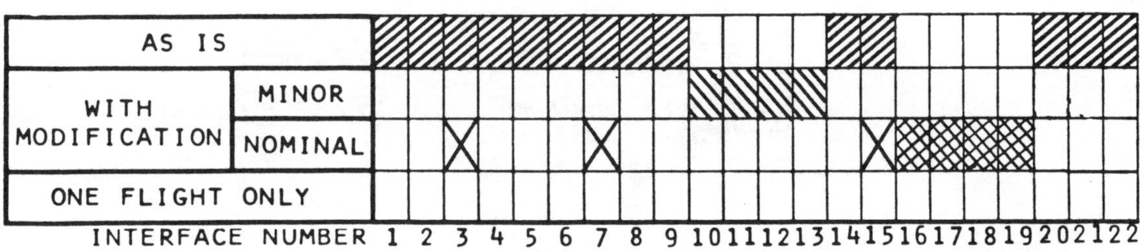

Figure 3-3. Example No. 1: Estimated Extended-Life Potential

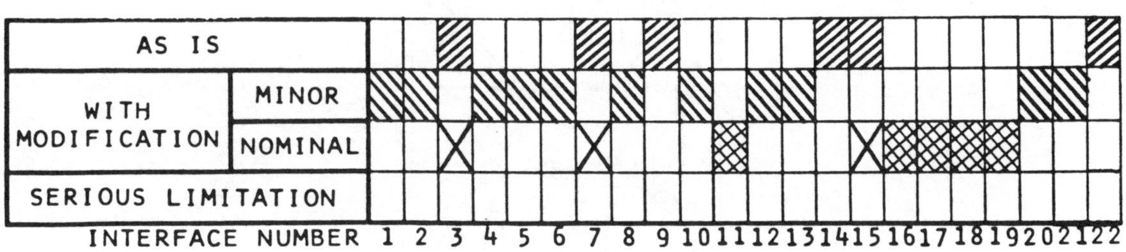

Figure 3-4. Example No. 1: Estimated Prolonged-Life Potential

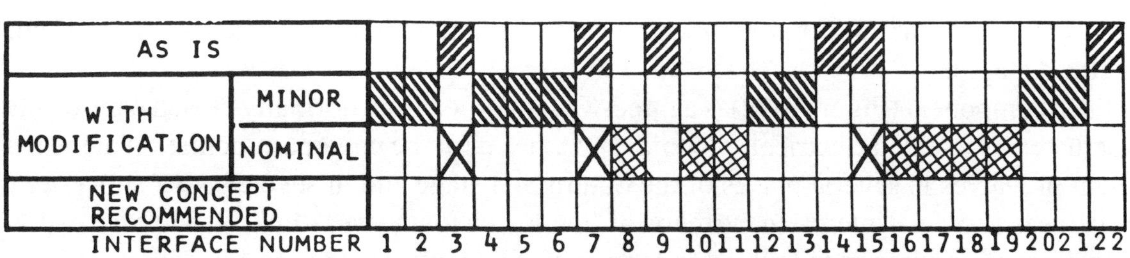

Figure 3-5. Example No. 1: Estimated Long-Life Potential

Note: ☒ Non-Interface recomendations

Control No.: J-2/208-1, J-2 Engine, Four Inch Normally Closed Main Fuel Valve, Rocketdyne P/N 409920X7; S/N 4082852 (Component Data Book Supplement Number 2)

Test Exposure:	73 J-2 Engine Test Firings
Test Cycles:	292 Cycles, based on 73 Test Firings, 4 Cycles Per Firing (146 @ Amb.; 146 @ LH_2)
Cycle Requirements:	1308 Cycles, Based On Qualification Requirements (860 @ Amb.; 346 @ LH_2; 102 @ +140°F)
Vibration Exposure:	156.2 Minutes

NOTE: In the Interface Location Drawings below, the numbers in circles are interface numbers.

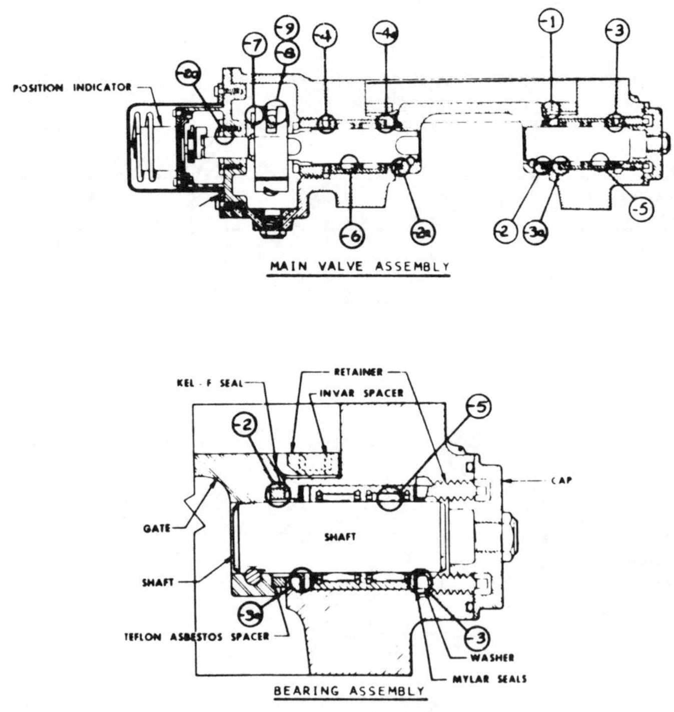

Figure 3-6. Wear Analysis Example No. 2: 4-in. J-2 Engine Main Fuel Valve

fuel valve of Rocketdyne's J-2 engine which powered the Saturn S-II (second stage) and the Saturn S-IVB (third stage).

The third component (Figure 3-12) is a three-way, two-position, normally closed control valve that is actuated by a sealed electrical solenoid. It was used for the pneumatic actuation of the main propellant prevalves in several places of the Saturn S-II stage and in several other subsystems.

This valve is of special interest in the realm of "the search for the weak link." It had been exposed earlier to planned endurance and vibration tests, which had resulted in specific recommendations for modifications to extend its life. These recommendations, subsequently implemented, included

NOTE: In the Interface Location Drawings below, the numbers in circles are interface numbers.

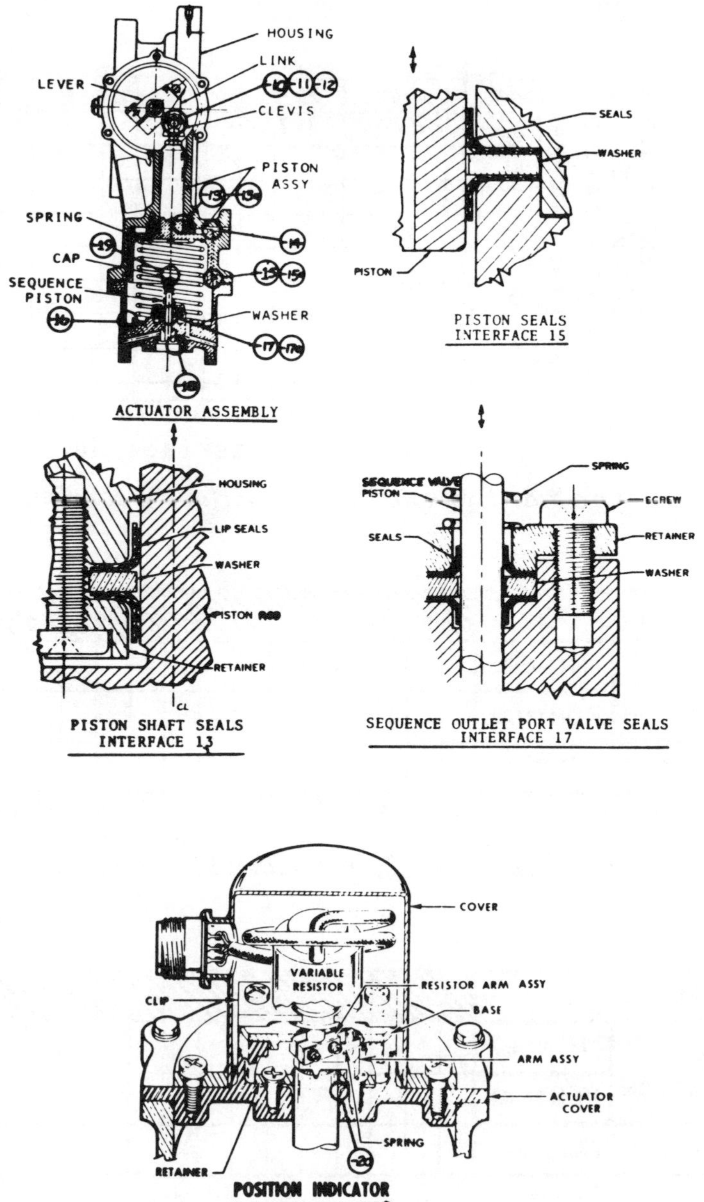

Figure 3-7. Wear Analysis Example No. 2: 4-in. J-2 Engine Main Fuel Valve, Interface Details

improved surface treatments for the solenoid armature guides (bushings and armature) and several material and dimensional changes at other interfaces. A valve, modified accordingly, was then subjected to renewed endurance tests. Figures 3-13, 3-14, and Table 3-3 summarize the new analysis results.

The figures and the table now register only relatively minor concerns and modifications. Especially interesting is the fact that the note accompanying Figure 3-14 points out a limitation (an undersized solenoid coil) which is unrelated to wear. If wear-related causes had not been eliminated first, as steps in the search for the weak links, it is probable that this shortcoming would

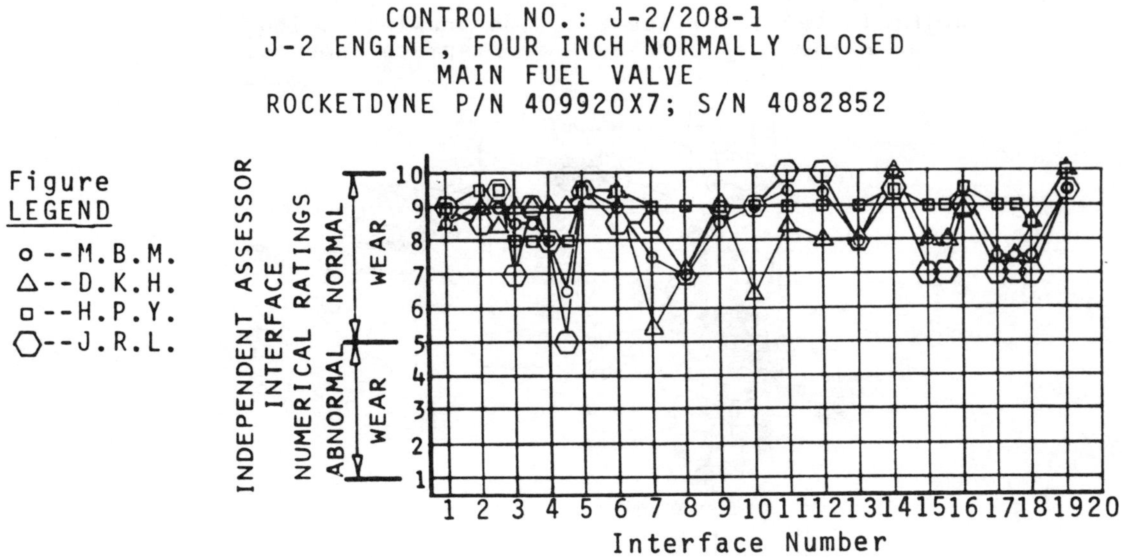

Figure 3-8. Example No. 2: Interface Wear Rating

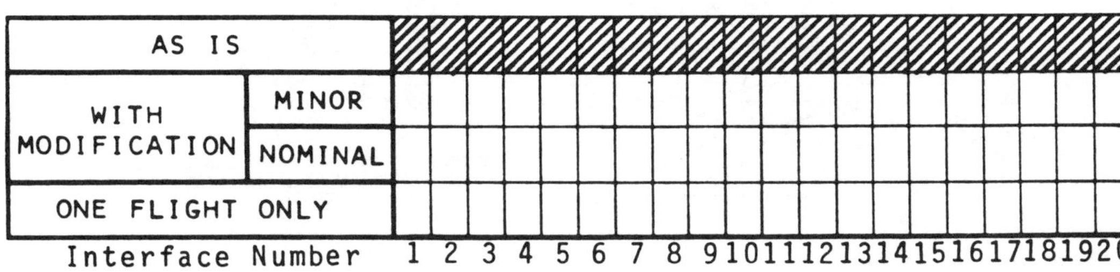

Figure 3-9. Example No. 2: Estimated Extended-Life Potential

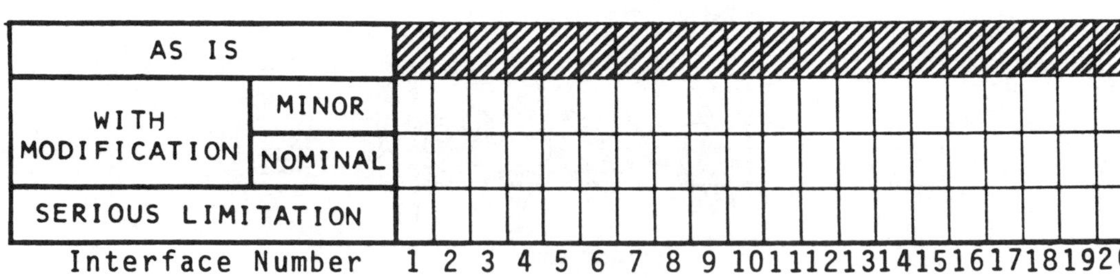

Figure 3-10. Example No. 2: Estimated Prolonged-Life Potential

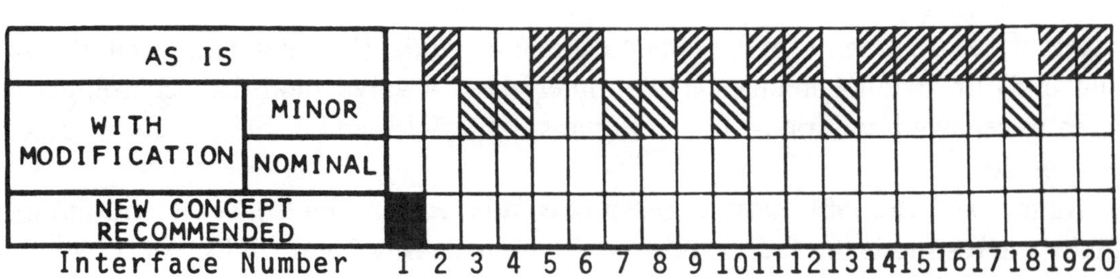

Figure 3-11. Example No. 2: Estimated Long-Life Potential

Control No.: S-II/104M1-2, Saturn S-II, 3/8-Inch Solenoid Valve Assembly, P/N ME284-0356-Modified (J. C. Carter P/N 6969-Modified)

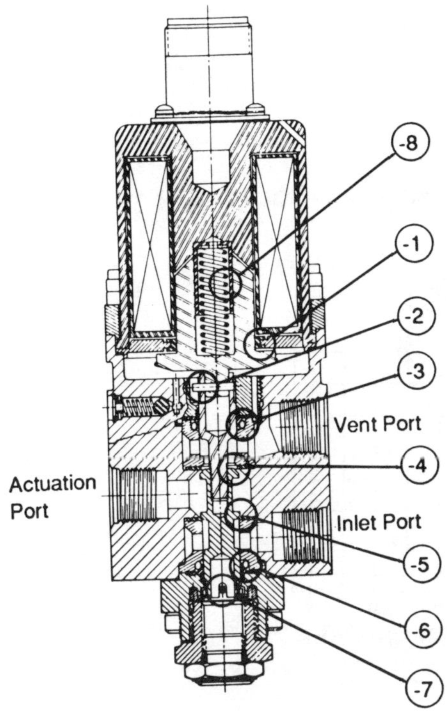

Test Exposure:	Endurance Tests
Test Cycles:	10,125 Ambient and Cryogenic Cycles
Cycle Requirement (Design Life):	2,000 Total (ambient and cryogenic) Cycles
Vibration Exposure:	600 Minutes Total, Random Vib. only (234 minutes in tangential axis and 366 minutes in longitudinal axis)
Vibration Requirement:	126 Minutes Total (18 minutes sinusoidal and 24 minutes random in each of 3 axes)

Figure 3-12. Wear Analysis Example No. 3: 3/8-in. Solenoid Valve

not have been clearly identified as a cause of leakage.

As we have illustrated by these three examples, the following steps in the wear-rating procedure should be taken.

After all analysis team members have made and submitted their assessments for each interface, the coordinator combines them into graphs (Figures 3-2, 3-8, and 3-13). In the graphs, each team member is identified by a symbol and by his initials. As was typical for more than a dozen other components analyzed in this manner, there always was very good agreement qualitatively, though somewhat less quantitatively, between the wear ratings of the component interfaces.

Following the distribution of the Rating Graphs, the team members meet to discuss and refine their assessments. If so indicated, they should consult a specialist before arriving at their final judgment of the component's life potential (Figures 3-3, 3-4, 3-5, 3-9, 3-10, 3-11, and 3-14). They will also generate a summary of the recommended modifications required to achieve the projected life potentials (Tables 3-1, 3-2, and 3-3). As may be seen, some of these recommendations include several for "Noninterfaces," which were deemed beneficial and noted during the discussions.

Based on these data, deciding whether or not to proceed with the modification(s) will be an Engineering Management decision; this will entail engineering drawings and fabrication of the modified parts. Also, a decision will have to be made on the scope of the verification tests needed to confirm the effectiveness of the changes.

During the study at Rockwell International's Space Division (see Reference 1), recommended changes based on wear-analysis ratings were actually incorporated, followed by verification tests,

CONTROL NO.: S-II/104M1-2
SATURN S-II 3/8-INCH SOLENOID VALVE
P/N ME284-0356-MODIFIED (J.C. CARTER P/N 6969-MODIFIED)

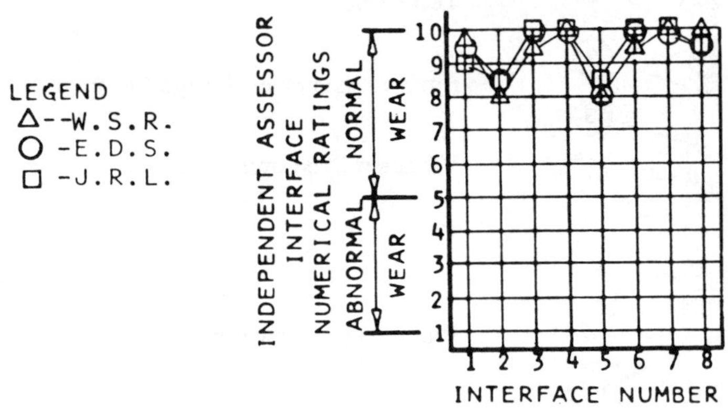

Figure 3-13. Example No. 3: Interface Wear Rating

EXTENDED LIFE, PROLONGED LIFE, AND LONG LIFE POTENTIALS

THE SOLENOID VALVE ASSEMBLY MODIFICATIONS HAVE DEMONSTRATED THEIR LONG LIFE POTENTIALS BY COMPLETING THE ENDURANCE TESTS AND EXPERIENCING MINIMAL WEAR. HOWEVER, SPORADIC DE-ENERGIZED LEAKAGE EXPERIENCED DURING THE ENDURANCE TEST, EVEN AFTER INCREASING THE DE-ENERGIZED SPRING PRELOAD, INDICATES A VALVE ASSEMBLY LIMITATION AS A LONG LIFE CANDIDATE. WHILE THE INTERFACES ARE MORE THAN ADEQUATE FOR LONG LIFE USAGE, THE VALVE IS SUSPECTED OF BEING LIMITED BY AN UNDERSIZED SOLENOID ASSEMBLY COIL.

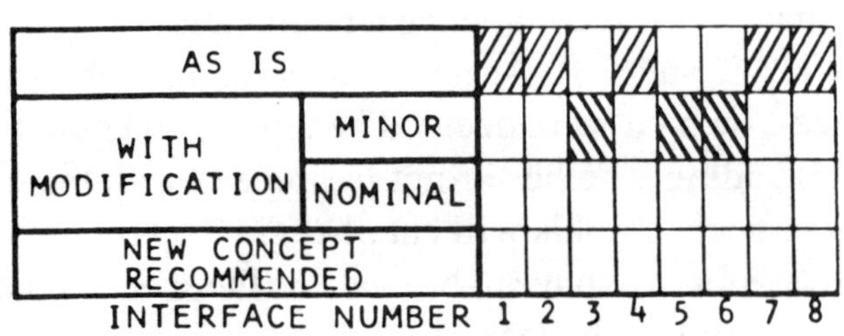

Figure 3-14. Example No. 3: Estimated Long-Life Potential

Table 3-1 Recommended modifications for increasing the operational life of the 1½-in. helium pressure regulator, Control No. S-IC/207-5 (see Fig. 3-1)

Interface number	Primary interface function	Recommendations for "Extended life"	Recommendations for "Prolonged life"	Recommendations for "Long life"
1	Main poppet seat	Suitable as is	Evaluate harder seal and poppet mat″ls. (e.g., in the 440C Cres hardness range)	
2	Main poppet to baffle gross leakage seals	Suitable as is	Investigate polishing baffle bore to a low-frequency high gloss finish, and alternate seal materials (e.g., BeCu)	
3	Ball screw ram ball return channel deflectors	☒ Noninterface recommendations: Investigate methods to improve alignment and retention of ball return channel deflectors on ball screw ram		
4 & 5	Ball screw assembly to housing support bearings	Suitable as is	Investigate a more positive press fit, axial loading of, or keying of unloaded bearing races to prevent slippage	
6 & 6a	Ball screw ram antirotational levers to housing	Suitable as is	Evaluate use of dry film lubed levers mating with a chrome plated 303Se Cres housing, or fabrication of levers from a more wear resistant material, BeCu	
7 & 7a	Ball screw ram antirotational levers (nonint.)	☒ Noninterface recommendations: Add thrust bearing washer between levers, shoulders to lever bushings, or recess lever faces back from the bearing inserts		
8	Ball screw ram to antirotational levers pin joint	Suitable as is	Investigate other surface finishes & treatments on pin and in ball screw ram pin bores	Evaluate addition of bearing inserts (e.g., BeCu) in ball screw ram pin bores
10	Ball screw nut to power bellows drive linkage pin	Evaluate fabrication of link bushing from a more adhesive wear resistant material, such as BeCu		Evaluate addition of self aligning BeCu spherical bearing in link
11	Ball screw nut to power bellows drive linkage pin joint	Evaluate use of thin dense chrome plated pin	Investigate addition of bearing inserts in ball screw nut bores, or keying pin to prevent rotation in ball screw nut bores	
12 & 13	Power bellows drive linkage stabilizer pin joint	Evaluate other pin, link, and lever materials and surface treatments (e.g., thin dense chrome plated pin in dry film lubricated link and lever bores, or in a link or lever fabricated from BeCu)		
15	Power bellows drive linkage stabilizer to body pin joint	☒ Noninterface recommendations: Use a dry film lubed lever, add thrust bearing washers on either side of lever, or shrink bushings in body pin bores to absorb lever side loads and prevent galling with body		
16, 17, 18, & 19	Tank and outlet checkout shuttle valve seats	Evaluate the use of separate checkout shuttle valve units, not integral with the regulator; for ground checkout purposes, the units could be connected to the regulator thru quick disconnects or existing flange mounts		
20 & 21	Pilot valve and flow limiter seats	Suitable as is	Evaluate the use of harder interfacing materials (e.g., press and retain a "stellite" seat into the control block)	

with excellent results. For instance, a large and complex (7 in.) vent and relief valve that was produced by the Ametec-Calmec Company and that had a design life of 600 cycles (under cryogenic conditions) successfully passed a 10,400-cycle endurance test, after selected modifications had been applied. A smaller valve, a hydrogen recirculation valve, also passed 10,000-cycle endurance tests, followed by 10 h of random vibration tests—thus demonstrating the effectiveness of recommended modifications when they are implemented. About half of all tests were conducted at cryogenic temperatures. All tests included periodic leak tests at selected seals. Subsequent wear analysis showed significantly reduced wear.

Table 3-2 Recommended modifications for increasing the operational life of the 4-in. J-2 engine main fuel valve, Control No. J-2/208-1 (see Figs. 3-6 and 3-7)

Interface number	Primary interface function	Recommendations for		
		"Extended life"	"Prolonged life"	"Long life"
1	Main gate seal	Suitable as is	Suitable as is	This gate sealing concept is not recommended for long life
3, 3a, 4, & 4a	Idler and drive shaft seals	Suitable as is	Suitable as is	Investigate new shaft materials such as BeCu and improve surface finishes. Future designs should attempt to eliminate these shaft seals
7	Lever pin to lever	Suitable as is	Suitable as is	Investigate improving alignment between lever and link to improve lever to link loading
8 & 10	Link to lever pin and clevis pin	Suitable as is	Suitable as is	Investigate self aligning feature for link, utilizing spherical bearing in the link bores
13 & 13a	Piston shaft seals	Suitable as is	Suitable as is	Investigate staggered seal arrangement. Also investigate a method to control contamination
15 & 15a	Piston seals	Suitable as is	Suitable as is	Control of contaminants will enhance the operational life. Investigate modification to incorporate a bellows type actuator
18	Sequence valve poppet seat	Suitable as is	Suitable as is	Improve seat design and investigate material selection of harder interfacing materials and improving poppet guide to provide more stability

Table 3-3 Recommended modifications for increasing the operational life of the 3/8-in. solenoid valve assembly, Control No. S-II/104M1-2 (see Fig. 3-12)

Interface number	Primary interface function	Recommendations for		
		"Extended life"	"Prolonged life"	"Long life"
3 & 6	Upper and lower poppet seal	Suitable as is	Suitable as is	Investigate 24281 seal to 24381 ring interface to determine cause of excessive scuffing and remedy situation
5	Main seat	Suitable as is	Suitable as is	Change seat material to Vespel as per interface 4; investigate and resolve inadequate de-energized loading problems and temperature sensitivity

Materials Aging

In Chapter 2, "Causes of Diminished Component Life," we briefly touched upon Materials Aging as the second major mechanism causing component degradation. While *Wear* was defined as the function of component usage or operational "cycles," *Materials Aging* was identified as a function of exposure to various environments and of elapsed time.

The selection of the most suitable material for the piece parts of a component is an important step in its design and usually involves metals and nonmetallic materials. As we pointed out in Chapter 2, both types are subject to wear, but it is widely agreed that metals are of no concern with respect to aging, as defined, for the time spans and environments encountered—even by Long-Life components. Although some metals may be subject to corrosion, this effect will be attributed to causes other than aging.

Thus, the concern with aging is predominantly with nonmetallic materials.

Age Life of Common Materials

In the late 1960s, those concerned at NASA and in the industry realized that revised NASA Apollo launch schedules would keep a number of command and service modules (CSMs) in storage longer than anticipated. As a result, the supplier of the CSM, Rockwell International's Space Division in Downey, California, was contracted to analyze all CSM components for their potential capability of age-life extension or need for replacement.

The results of this study are extremely valuable and are presented here in tabulated form as Table 3-4. Note that the acronym "NR" used in the "References" column stands for North-American Rockwell Corp., now Rockwell International, Inc.

The table is largely self-explanatory. However, the following comments should be made: The stated life spans assume storage at room temperature (Earth environment), without exposure to effects for which the material is not suitable. Recommendations and limitations for other conditions are listed in applicable columns of the table.

As we mentioned, the data presented in Table 3-4 had been generated for a specific purpose, i.e., the assurance that the spacecraft and its components, which had been stored longer than planned, were still safe to use. However, in this process, much data was accumulated which has value well beyond the original need. More than likely, though, the table cannot answer all questions arising in connection with materials selections of the future. For instance, applications involving highly corrosive fluids may require special investigations or specified servicing procedures. Nevertheless, the information contained in the table will be highly applicable to a large portion of future space hardware—notably cryogenic systems, pneumatic and pressurized-gas systems—and a multitude of other applications similar to the Apollo CSM and Saturn vehicles.

The table does not reflect the effects of radiation. Published reports appear now to be in general agreement that we should not be concerned about the cumulative *natural* radiation doses during space missions in the foreseeable future. A possible exception is the use of materials in the vicinity of radioactive sources, such as a radioisotope thermoelectric generator (RTG) or a nuclear engine. In such cases, these special conditions must be considered when materials are being selected. NASA George C. Marshall Space Flight Center reported upon the results of a study conducted by General Dynamics, during which Saturn V components were evaluated for potential use in

Table 3-4 Room temperature (Earth environment) storage life of common materials used in spacecraft and launch vehicles (Sheet 1)

ITEM	COMMON NAME	TRADE NAME	ACRONYMS	CHEMICAL NAME	CLASS	FAMILY	SPECIFIC WEIGHT	AGE LIFE (YRS.)		PRINCIPAL USES
								DOCU-MENTED	MIN. PRE-DICTED	
1	SILICON RUBBER	SILASTIC (21) SI-D-FLEX (24)	PSI	POLYSILOXANE	ELASTOMER	SILICONE	1.1 – 1.6	10	20	SEALS FOR USE AT TEMPEATURES FROM −100 F TO 500 F
2	FLUOROSILICONE RUBBER	SILASTIC	FVS	FLUORO-VINYL SILANE	ELASTOMER	FLUOROSILICONE	1.4	10	20	HIGH TEMPERATURE AND OIL RESISTANT SEALS AND GASKETS
3	EPR RUBBER	EPCAR EPSYN-ROYALENE	EPR	ETHYLENE PROPYLENE RUBBER	ELASTOMER	ETHYLENE PROPYLENE	.86	10	20	GENERAL PURPOSE MECHANICAL GOODS, OZONE AND HYDRAZINE RESISTANT
4	SBR RUBBER	PLIOFLEX AMERIPOL SOLPRENE	BUNA-S GR-S	STYRENE-BUTADIENE RUBBER	ELASTOMER	STYRENE-BUTADIENE	.94	10	20	GASKETS, SEALS, O-RINGS – (−60 F TO 180 F)
5	POLYACRYLATE RUBBER	CYANACRYL HYCAR THIACRIL	ACM	POLYACRYLATE RUBBER	ELASTOMER	ACRYLIC	1.09	10	15	GASKETS, SEALS, O-RINGS – (−60 F TO 175 F) – ALSO USED AS SEALANT
6	BUTYL RUBBER	BUCAR BUTYL ENJAY	IIR, BUTYL, GR-1	ISOBUTYLENE-ISOPRENE RUBBER	ELASTOMER	ISOBUTYLENE-ISOPRENE	.92	10	20	INSULATION, VIBRATION DAMPENERS, INNER TUBES
7	NITRILE RUBBER	HYCAR KRYNAC CHEMIGUM	KBR, BUNA-N	BUTADIENE ACRYLONITRILE RUBBER	ELASTOMER	ACRYLONITRILE	1.09	10	15	SEALS FOR USE WITH OILS
8	HYPALON	HYPALON	CSN	CHLOROSULFONATED POLYETHYLENE RUBBER	ELASTOMER	CHLOROSULFONATED POLYETHYLENE	1.18	10	20	CHEMICAL RESISTANT DIAPHRAGMS – OZONE RESISTANT
9	NEOPRENE	NEOPRENE	CR	CHLOROPRENE RUBBER	ELASTOMER	CHLOROPRENE	1.25	10	20	GENERAL PURPOSE WEATHER AND OZONE EXPOSURE SEALS, ETC.
10	NITROSO RUBBER	RESISTAZINE -88	CNR	CARBOXY-NITROSO RUBBER	ELASTOMER	CARBOXY-NITROSO	–	4–5	4–5	SEALS FOR CONTACT WITH N_2O_4 OXIDIZER
11	FLUORINATED RUBBER	VITON FLUOREL	FPM	FLUOROCARBON	ELASTOMER	FLUORINATED HYDROCARBON	1.86	10	20	SEALS FOR 450 F AND/OR IN CONTACT WITH FUELS
12	SYNTHETIC RUBBER	AMERIPOL CORAL NATSYN	IR	SYNTHETIC POLYISOPRENE RUBBER	ELASTOMER	ISOPRENE	.91	6	6	SHOCKS, MOUNTS, BUSHINGS, WATER/GLYCOL BLADDERS
13	THIOKOL RUBBER	THIOKOL RUBBER	PTR	POLYSULFIDE RUBBER	ELASTOMER	POLUSULFIDE	1.35	10	20	WEATHER AND OIL RESISTANT SEALS, GASKETS, CALKING COMPOUNDS
14	URETHANE RUBBER	–	AU	POLYESTER URETHANE RUBBER	ELASTOMER	POLYURETHANE	1.25	6	20	ELECTRICAL POTTING, BAGS, ROLLERS, BUMPERS AND GREASE SEALS
15	URETHANE RUBBER	ADIPRENE	EU	POLYETHER URETHANE RUBBER	ELASTOMER	POLYURETHANE	1.12 – 1.5	10	20	ELECTRICAL POTTING, GREASE SEALS, ETC.

Table 3-4 Room temperature (Earth environment) storage life of common materials used in spacecraft and launch vehicles (Sheet 1 continued)

ITEM	REMARKS	SERVICE ENVIRONMENT LIMITATIONS	REFERENCES	STABILITY IN SPACE	SPACE ENVIRONMENT COMMENTS
1	SEALS VALVES, ETC., FOR 11 YEARS WITHOUT FAILURE IN ACTUAL TESTS. SEALS AFTER ACCELERATED AGING.	EXPOSURE TO MOST ORGANIC LIQUIDS – (FLUOROSILICONE ELASTOMERS ARE EXCEPTION)	(1) MFSC REPORT IN ASTN–M–71–1 DATED 2–2–71 PP. 26–27 (2) MIL–HDBK–695 (3) MSFC–STD–105 (4) NR TR 332359	(a, b)	(a) CRAZES AND CRACKS WHEN EXPOSED TO CONTINUOUS CRYO–TEMPERATURES DURING QUIESCENT STORAGE, EVEN UNDER ZERO STRESS, IF NOT RIGIDLY RETAINED. CYCLING TO WARMER SUBZERO TEMPERATURES ACCENTUATES THE PROBLEM. (b) NO EVIDENCE OF DEGRADATION (PERIODS UP TO 1 YEAR AT N B STDS) WHEN RIGIDLY CONFINED. (c) SILICONE SEALANTS, LUBRICANTS, AND POTTING COMPOUNDS EMIT CONDENSIBLE OILS AND CONSTITUENTS UNDER HARD VACUUM (10^{-4} TORR) AT TEMPERATURES UP 200 F DURING CYCLIC EXPOSURE AND QUIESCENT STORAGE. LESS THAN 1/10 PERCENT WEIGHT LOSS AT ROOM TEMPERATURE OR SUB–ZERO IN HARD VACUUM. (d) NO EVIDENCE OF VCM PRODUCTS OR EXCESSIVE WEIGHT LOSS IN VITON WHEN EXPOSED, UNCONFINED, IN HARD VACUUM AT TEMPERATURES FROM –100 TO 450 F. (e) DEGRADES IN (10^{-6} TORR) VACUUM AND 122 F TEMPERATURE CAUSED BY UV EXPOSURE. EVIDENCE OF VCM DEPOSITS. (f) SOME WEIGHT LOSS AND VCM AT ROOM TEMPERATURE (10^{-6} TORR) VACUUM. (g) POLYSULFIDE EXPERIENCES CONSIDERABLE WEIGHT LOSS AT ROOM TEMPERATURE IN HARD VACUUM. FIFTY PERCENT GREATER WEIGHT LOSS AT 250 F IN HARD VACUUM. PASTE FORM GREATEST WEIGHT LOSS. (h) WEIGHT LOSS AND/OR VCM UP TO 1/2 PERCENT IN HARD VACUUM (10^{-5} TORR) AT 250 F, UP TO 1/4 PERCENT AT ROOM TEMPERATURE IN (10^{-5} TORR) VACUUM. (j) PREFERRED SEAL MATERIALS FOR LONG TERM SERVICE WITH HIGH ENERGY FUELS OR OXIDIZERS. SERVICE TEMPERATURES –300 F TO 700 F. (k) MAY BE USED FOR HYDRAULIC AND PNEUMATIC SEALS (–300 F TO 700 F) (l) SECTION 12, REFERENCE 1: SUPPLEMENT 19 TO REPORT SD71–267 (TABLE 4.6–1). (m) HIGHER RADIATION DAMAGE THRESHOLD TOLERANCE. NO VCM. (n) MODERATE RADIATION DAMAGE THRESHOLD TOLERANCE. NO VCM. (o) NOT RATED FOR RADIATION DAMAGE. THRESHOLD SHOULD COMPARE WITH PHENOLIC. NO VCM. (p) LOW RADIATION DAMAGE THRESHOLD TOLERANCE, HOWEVER, SHOULD BE ADEQUATE FOR LOWER STRESSED APPLICATIONS. (q) NO EVIDENCE OF CRITICAL MECHANICAL DAMAGE IN SPACE ENVIRONMENTS.
2	SEALS VALVES, ETC., FOR 11.5 YEARS WITHOUT FAILURE IN ACTUAL TESTS. ACCELERATED HIGH TEMPERATURE AGING SHOWS MINIMAL HARDNESS CHANGE.	LOW MECHANICAL PROPERTIES	(1) IBID, PP. 24–25 (10) DOW CORNING BULLETIN 09–285, AUGUST 1965	(l)	
3	SEALS EXPERIENCED NO FAILURES AFTER TESTING FOR NINE YEARS.	AVOID OIL AND FUEL EXPOSURE, SUB–ZERO AND HIGH TEMPERATURES.	(1) IBID, PP. 48–49 (2) IBID	(l)	
4	INSTALLED AS SEALS (11 YEARS WITH NO LEAKS) – FAIR ACID RESISTANCE.	POOR RESISTANCE TO HYDROCARBONS – POOR SUNLIGHT AGING.	(1) IBID, PP. 38–39 (14) IBID, P. 353	(f, l)	
5	EXCELLENT OXIDATION RESISTANCE, FAIR ACID RESISTANCE, AGING AT 90 RH AT 200 F ≤12 – 15 YEARS AT R. T.	POOR RESISTANCE TO AROMATIC HYDROCARBONS, OIL AND GASOLINE.	(40) NAFI REPORT TR 1178 (14) IBID, P. 354	(h)	
6	SEALS VALVES, ETC., > 11 YEARS ACCELERATED AGING TESTS OF SEALS WITHOUT FAILURE.	AVOID EXCESSIVE TEMPERATURE AND EXPOSURE TO HYDROCARBONS AND FUELS.	(1) IBID, PP. 22–23 (4) IBID	(l)	
7	SEALS VALVES, ETC., FOR 11 YEARS WITH 3 PERCENT FAILURES IN ACTUAL TEST. SEALS AFTER ACCELERATED AGING INSTALLED AFTER 9 YEARS SHELF STORAGE AND OPERATED SATISFACTORILY.	AVOID OUTDOOR WEATHERING AND EXPOSURE TO EXTREME HIGH TEMPERATURES.	(1) IBID, PP. 36–37, 46–47 (4) IBID (7) IBID	(l)	
8	NO SEAL FAILURES EVEN AFTER 10 YEARS OF TESTING	AVOID IMMERSION IN FUELS AND EXPOSURE TO EXTREME HIGH TEMPERATURES.	(1) IBID, PP. 28–29	(f, l)	
9	SEALS VALVES, ETC., FOR 11 YEARS WITHOUT FAILURE IN ACTUAL TESTS. NEOPRENE ADHESIVE IMPROVES AFTER ACCELERATED AGING.	AVOID TEMPERATURES ABOVE 212 F.	(1) IBID, PP. 34–35 (4) IBID	(l)	
10	INSPECTION OF 4–1/2 YEAR OLD VALVE SHOWED NO DETERIORATION.	COSTLY . . AVOID OTHER DESIGN APPLICATIONS.	(6) NR TR 302017 (4) IBID	-	
11	FLAME RESISTANT. SEALS VALVES, ETC., FOR 11 YEARS WITHOUT FAILURE IN ACTUAL TESTS. INSTALLED AFTER 9 YEARS SHELF STORAGE AND OPERATED SATISFACTORILY. SEALS AFTER ACCELERATED AGING.	DO NOT USE AT TEMPERATURES BELOW –10 F.	(1) IBID, PP. 44–45 (7) NR/RD REPORT MPR71–647 APRIL 7, 1971, TABLES IV AND V (4) IBID	(d)	
12	SEALS AFTER 11.5 YEARS IN ACTUAL TESTS WITH ABOUT 6 PERCENT FAILURE. SEALS AFTER ACCELERATED AGING.	AVOID OUTDOOR EXPOSURE AND EXPOSURE TO OILS AND FUELS.	(1) IBID, PP. 32–33 (4) IBID, ALSO LR9190–4148 (11) AIRESEARCH REPORT SS–1984–R DATED 10–23–70	(n, t)	
13	SEALS VALVES, ETC., FOR 9.5 YEARS IN ACTUAL TESTS WITH ABOUT 3 PERCENT FAILURE.	HAS GENERALLY LOW MECHANICAL PROPERTIES. AVOID TEMPERTURES > 180 F.	(1) IBID, PP. 40–41	(g)	
14	ONE TYPE OF POLYESTER REVERTS IN 6 YEARS, OTHERS LAST UP TO 10 YEARS – VERIFY TYPES.	AVOID MOISTURE EXPOSURE, HIGH TEMPERATURS, ACIDS AND ALKALIS.	(1) IBID, PP. 42–43 (8) NAVAL AVIONICS FACILITY/ INDIANAPOLIS REPORT NAFI–TR–1372, 2–27–69	(l)	
15	LESS PRONE TO REVERSION THAN POLYESTER TYPES.	AVOID MOISTURE EXPOSURE, HIGH TEMPERTURES, ACIDS AND ALKALIS.	(8) IBID	(l)	

(continued on next page)

Table 3-4 Room temperature (Earth environment) storage life of common materials used in spacecraft and launch vehicles (Sheet 2)

ITEM	COMMON NAME	TRADE NAME	ACRONYMS	CHEMICAL NAME	CLASS	FAMILY	SPECIFIC WEIGHT	AGE LIFE (YRS.)		PRINCIPAL USES
								DOCU-MENTED	MIN. PRE-DICTED	
16	RTV SILICON SEALANT	RTV. SILASTIC	RTV	POLYSILOXANE	ELASTOMETRIC SEALANT	SILICONE	1.12 – 1.5	10	15	POTTING, SEALING, ADHESIVE BONDING FROM –100 F TO 500 F
17	EPOXY PLASTIC	–	–	EPOXIDE	THERMOSETTING PLASTIC	EPOXIDE	1.1 – 1.8	10	20	POTTING, ADHESIVES, MOLDINGS AND STRUCTURAL LAMINATES
18	MELAMINE	MELMAC CYMEL	–	MELAMINE–FORMALDEHYDE	THERMOSETTING PLASTIC	AMINO	1.43 – 2.0	10	20	ELECTRICAL MOLDINGS – FLAME RESISTANT LAMINATES
19	PHENOLIC PLASTIC	BAKELITE	–	PHENOL–FORMALDEHYDE	THERMOSETTING PLASTIC	PHENOLIC	1.24 – 3.0	10	20	STRUCTURAL LAMINATES AND MOLDED PARTS
20	PBI PLASTIC	–	PBI	POLYBENZIMIDAZOLE	THERMOSETTING PLASTIC	POLYBENZIMIDAZOLE	1.2 – 1.5	10	20	LAMINATES SUBJECT TO TEMPERATURES UP TO 600 F
21	POLYESTER PLASTIC	–	–	ALKYD	THERMOSETTING PLASTIC	ALKYD	1.1 – 2.0	10	20	COVERS, HOUSINGS, PANELS AND MOLDED PARTS
22	POLYIMIDE PLASTIC	VESPEL	PI	POLYIMIDE	THERMOSETTING PLASTIC	AROMATIC IMIDE	1.41	10	20	WIRE INSULATION, STRUCTURAL LAMINATES, MOLDINGS FOR HIGH TEMPERATURE USE
23	SILICONE PLASTIC	–	–	PHENYL SILICONE	THERMOSETTING PLASTIC	SILICONE	1.75 – 2.0	10	–	THERMAL, MOISTURE AND OXIDATION RESISTANT LAMINATES AND MOLDINGS
24	ACETAL PLASTIC	DELRIN CELCON	–	POLYACETAL	THERMOPLASTIC PLASTIC	POLYACETAL	1.41 – 1.61	10	–	INJECTION MOLDED PARTS – BUSHINGS, GEARS, ETC., UP TO 200 F
25	DIAL PLASTIC	DIAL	DAP	DIALLYL PHTHALATE	THERMOSETTING PLASTIC	POLYESTER	1.5 – 1.8	10	20	ELECTRICAL MOLDINGS, STRUCTURAL LAMINATES
26	KEL–F PLASTIC	KEL–F	PCTFE	POLYCHLOROTRI–FLUOROETHYLENE	THERMOPLASTIC PLASTIC	FLUOROCARBON	2.10 – 2.15	10	20	CHEMICAL AND HIGH TEMPERATURE RESISTANT SEALS, MOLDINGS AND FILM
27	NYLON	NYLON	–	POLYAMIDE	THERMOPLASTIC PLASTIC	AMIDE	1.0 – 1.4	10	10	CHEMICAL RESISTANT SEALS, GEARS, BEARINGS, MOLDINGS AND ELECTRICAL INSULATION
28	POLYETHYLENE POLYPROPYLENE	POLYTHENE	PE	POLYETHYLENE POLYPROPYLENE	THERMOPLASTIC PLASTIC	POLYOLEFIN	.91 – 1.2	10	20	ELECTRICAL INSULATION, FILMS, MOLDINGS, PACKAGING
29	POLYCARBONATE PLASTIC	MERLON LEXAN	–	POLYPHENYL–CARBONATE	THERMOPLASTIC PLASTIC	POLYCARBONATE	1.1 – 1.2	10	20	HIGH IMPACT STRENGTH TRANSPARENCIES, MOLDINGS
30	ACRYLIC PLASTIC	PLEXIGLASS LUCITE	–	POLYMETHYL METHACRYLATE	THERMOPLASTIC PLASTIC	ACRYLIC	1.1 – 1.2	10	20	MOLDINGS, HIGH QUALITY OPTICAL TRANSPARENCIES

Table 3-4 Room temperature (Earth environment) storage life of common materials used in spacecraft and launch vehicles (Sheet 2 continued)

ITEM	REMARKS	SERVICE ENVIRONMENT LIMITATIONS	REFERENCES	STABILITY IN SPACE	SPACE ENVIRONMENT COMMENTS
16	INSPECTION OF 6 YEAR OLD SWITCHES SHOWS NO DETERIORATION. TEN YEARS EXPERIENCE ON X-15. NO FAILURES. SEALS AFTER ACCELERATED AGING.	AVOID EXPOSURE TO MOST ORGANIC BASE LIQUIDS – SUBJECT TO GLASS TRANSITION AT TEMPERATURES UP TO –150 F.	(2) Ibid (5) DOW CORNING BULLETINS 08–418, 08–431	(c, e)	(r) LOW RADIATION DAMAGE TOLERANCE AND UNSATISFACTORY VCM (EXPOSED). (s) WHEN USED AS PART OF A COMPLETE SYSTEM – NOT EXPOSED TO SPACE ENVIRONMENT – THE AGE LIFE IS COMMENSURATE WITH THAT OF THE COMPLETE SYSTEM. (t) STRENGTH FALLS OFF WHEN EXPOSED DIRECTLY TO SOLAR UV. (u) NO CURRENT RATING FOR SPACE ENVIRONMENT EXPOSURE. (v) RADIATION EXPOSURE DEGRADES ELECTRICAL INSULATION PROPERTIES. (w) WEIGHT LOSS COMPARABLE TO THAT OF NYLON AT TEMPERATURE UP TO 200 F. (x) OPTICAL PROPERTIES DEGRADE WITH PROLONGED EXPOSURE TO SOLAR RADIATION AND/OR UV.
17	NO CHANGE AFTER 3 YEARS, STRESSED AND UNSTRESSED. SATISFACTORY 11 YEAR MILITARY AIRCRAFT USAGE.	AVOID TEMPERATURES ABOVE 300 F – 350 F. GENERALLY POOR FLAME RESISTANCE.	(12) ASD–TR61–145, PP. 16–18	(n)	
18	LONG EXPERIENCE IN USE IN COMMERCIAL AND AEROSPACE APPLICATIONS WITH LITTLE HISTORY OF FAILURE.	AVOID TEMPERATURES ABOVE 350 F CONTINUOUS.	(15) MODERN PLASTICS ENCY. 1969–70, P. 104 (16) 1971 MTLS. ENGRG. MTL. SELECTOR, PP. 212–213	(n, u)	
19	GOOD ABLATOR BINDER. NO CHANGE AFTER 3 YEARS, STRESSED AND UNSTRESSED. PARTS 15 AND 30 YEARS OLD SHOW NO DETERIORATION.	AVOID TEMPERATURES ABOVE 450 F – 500 F. GENERALLY POOR FLAME RESISTANCE.	(12) IBID, PP. 16–19	(m)	
20	GOOD CRYOGENIC PROPERTIES. COSTLY. USE EPOXY OR SILICONE FOR LOWER TEMPERATURE APPLICATIONS.	REQUIRES HIGHER CURE TEMPERATURE THAN PI.	(43) MODERN PLASTICS ENCY. 1968069, PP. 106–199	(o)	
21	SUBJECT TO GRADUAL CRAZING WITH LONG EXPOSURE TO SUNLIGHT.	POOR RESISTANCE TO STRONG ALKALIS WHEN GLASS REINFORCED.	(14) IBID, PP. 229–231 AND 241	(p)	
22	RETAINS USEFUL STRENGTH UP TO 40,000 HOURS AT 500 F. FIRE RESISTANT IN O_2 ATMOSPHERE.	COSTLY – HIGH PROCESSING TEMPERATURES. DIFFICULT TO CONTROL LAMINATE POROSITY.	(13) DuPONT BULLETIN A–53349 "VESPEL" FEBRUARY 1967 (14) "POLYIMIDE RESINS FOR AEROSPACE APPLICATIONS" 23RD SPI CONF. 2–6–63	(o)	
23	10,000 HOURS AGING AT 450 F (GOOD PROPS.) $\leq$10 YEAR AGING AT AMBIENT.	LOW INTERLAMINAR BOND STRENGTH.	(41) NR REPORT NA–60–964, DATED 07–28–60	(p)	
24	GOOD RESISTANCE TO ORGANIC SOLVENTS, HYDROCARBONS, ETC.	UNSATISFACTORY WITH MINERAL ACIDS AND OXIDIZING REAGENTS.	(43) IBID, PP. 115–117 (14) IBID, P. 258	(p, w)	
25	LONG COMMERCIAL AND AEROSPACE SERVICE WITH LITTLE HISTORY OF FAILURE.	TEMPERATURE EXPOSURES OVER 350 – 400 F CAUSE EXCESSIVE DEGRADATION.	(15) IBID, PP. 97–98, 970–971	(s, t)	
26	SEALS AFTER 11 YEARS IN ACTUAL TESTS WITHOUT LEAKING. LIP SEALS OPERATED SATISFACTORILY IN H–1 ENGINE AFTER 9 YEARS STORAGE.	AVOID CONTINUOUS TEMPERATURES > 400 F – SUBJECT TO COMPRESSION CREEP.	(1) IBID, PP. 48–49 (19) R/D REPORT, MPR 71–647, DATED 4–8–71	(n, q)	
27	SEALS AFTER 11 YEARS IN ACTUAL TESTS WITHOUT LEAKING. CREEP MODULUS DECREASES LESS THAN 20 PERCENT AFTER 2,000 HOURS AT 1,400 PSI.	AVOID MOISTURE AND SUNLIGHT UNLESS SPECIALLY FORMULATED.	(1) IBID, PP. 48–49 (15) IBID, P. 791	(r)	
28	SHOWS SATISFACTORY PROPERTIES AFTER ACCELERATED AGING EQUIVALENT TO 10 YEARS AT ROOM TEMPERATURE.	AVOID TEMPERATURES OVER 180 AND 200 F, RESPECTIVELY, OUTDOOR EXPOSURE WITHOUT ADDITIVES (STABILIZERS, ETC.)	(21) RAYCLAD LAB REPORT NO. 699, DATED 12–4–63, P. 13 (22) RAYCLAD SPECIFICATION RT–350, DATED 12–1–65	(n, t, v)	
29	CREEP MODULUS DECREASES ONLY 5 PERCENT FROM 10 HOURS TO 11,000 HOURS AT 160 F UNDER A LOAD OF 500 PSI.	AVOID TEMPERATURES > 325 F. POORER OPTICAL PROPERTIES THAN ACRYLICS.	(15) IBID, PP. 948–949, 979	(q, x)	
30	SATISFACTORY AFTER 15 YEARS OUTDOOR EXPOSURE.	AVOID TEMPERATURES > 240 F. POOR ABRASION RESISTANCE.	(15) IBID, P. 89	(t, x)	

(continued on next page)

Table 3-4 Room temperature (Earth environment) storage life of common materials used in spacecraft and launch vehicles (Sheet 3)

ITEM	COMMON NAME	TRADE NAME	ACRONYMS	CHEMICAL NAME	CLASS	FAMILY	SPECIFIC WEIGHT	AGE LIFE (YRS.) DOCU-MENTED	AGE LIFE (YRS.) MIN. PRE-DICTED	PRINCIPAL USES
31	TEFLON	TEFLON	TFE, FEP	TETRAFLUORO-ETHYLENE FLUORINATED ETHYLENE PROPYLENE	THERMOPLASTIC PLASTIC	FLUOROCARBON	2.1 – 2.3	10	20	HIGH TEMPERATURE–CHEMICAL RESISTANT SEALS, GASKETS, BLADDERS AND WIRE
32	KYNAR PLASTIC	KYNAR	PVF_2	POLYVINYLIDENE FLUORIDE	THERMOPLASTIC PLASTIC	FLUOROCARBON	1.77	10	20	HIGH TEMPERATURE – CHEMICAL RESISTANT FILM AND INSULATION
33	MYLAR FILM	MYLAR	–	POLYETHYLENE TEREPHTHALATE	THERMOPLASTIC PLASTIC	POLYESTER	1.39	10	20	TAPES, ELECTRICAL INSULATION, ALUMINIZED FILMS, PACKAGING
34	HT 424 ADHESIVE	HT 424	–	PHENOLIC MODIFIED EPOXIDE	STRUCTURAL ADHESIVE	EPOXY–PHENOLIC	–	10	20	STRUCTURAL METAL–TO–METAL AND SANDWICH BONDS
35	FECAL BAG ADHESIVE	X1140/X1160	–	(PROPRIETARY)	PRESSURE SENSITIVE ADHESIVE	SYNTHETIC RUBBER	–	5	10	FORMULATED TO BOND APOLLO FECAL BAG
36	PRIMER S	–	–	SILANES (AND OTHERS)	PRIMER	–	–	10	20	TO PROMOTE ADHESION AND PREVENT CORROSION WITH ADHESIVE SYSTEMS
37	FORMVAR COATING	FORMVAR	–	POLYVINYL FORMAL	WIRE COATING	–	–	10	20	PRIMARILY WIRE COATING UP TO 220 F CONTINUOUS DUTY
38	FIRE RETARDANT COATING	LADICOTE	–	(PROPRIETARY)	FIRE RETARDANT COATING	FILLED FLUOROCARBON ELASTOMER	–	10	20	FIRE RESISTANT OVERCOATING FOR MANNED SPACECRAFT APPLICATIONS
39	DACRON CLOTH	DACRON	–	POLYETHYLENE TEREPHTHALATE	SYNTHETIC FIBER	POLYESTER FIBER	1.38	10	20	OXIDATION AND CHEMICAL RESISTANT FABRICS
40	FISH PAPER	VULCANIZED FIBER	–	–	INSULATING PAPER	PHENOLIC IMPREGNATED CELLULOSE	–	10	–	HIGH DIELECTRIC WIRE RESISTANT FABRICS
41	NOMEX CLOTH	NOMEX	–	AROMATIC POLYAMIDE	SYNTHETIC FIBER	POLYAMIDE	1.20	10	15	MODERATE HEAT RESISTANT HONEYCOMB CORE, FABRIC, AND CLOTHING
42	ANDOK C GREASE	ANDOK C	–	–	GREASE	MINERAL OIL – SODIUM SOAP	–	10	–	GENERAL LUBE APPLICATIONS IN SEALED CONTAINERS OR AT AMBIENT CONDITIONS
43	DIESTER LUBRICANT	BEACON 325	–	–	GREASE	DIESTER	–	10	15	BEARING LUBRICANT FOR LONG LIFE AT MODERATE TEMPERATURES
44	KRYTOX LUBE	KRYTOX 240 AC	–	–	GREASE	FILLED FLUOROCARBON OIL	–	10	–	OXIDATION RESISTANT LUBRICANT – GOOD CHEMICAL RESISTANCE
45	SILICONE GREASE	DC 33 VERSILUBE S–300	–	–	GREASE	SILICONE FLUID – METAL SOAP	–	10	20	HIGH TEMPERATURE APPLICATIONS – MUST BE CONTAINED FOR SPACE VACUUM USE
46	DRY FILM LUBES	–	–	–	DRY LUBRICANT	ORGANIC BINDERS AND INORGANIC BINDERS	–	10	20	USED FOR SPACE APPLICATIONS WHERE GREASES AND OIL CANNOT BE CONTAINED.

Table 3-4 Room temperature (Earth environment) storage life of common materials used in spacecraft and launch vehicles (Sheet 3 continued)

ITEM	REMARKS	SERVICE ENVIRONMENT LIMITATIONS	REFERENCES	STABILITY IN SPACE	SPACE ENVIRONMENT COMMENTS
31	SEALS AFTER 11 YEARS IN ACTUAL TESTS WITHOUT LEAKING. UNAFFECTEDAFTER 10 YEARS WEATHER EXPOSURE.	AVOID APPLICATIONS WHERE 2–5 PERCENT CREEP CANNOT BE TOLERATED.	(1) IBID, PP. 48–49 (17) DuPONT CO. "TEFLON MECH. DESIGN DATA"	(n, q, r)	(SEE PRECEDING PAGES)
32	UNAFFECTED AFTER 8 YEARS EXPOSURE TO INDUSTRIAL ATMOSPHERE. CREEP 2 PERCENT AFTER 10,000 HOURS AT 500 PSI LOAD, ROOM TEMPERATURE.	AVOID CONTINUOUS EXPOSURE TO TEMPERATURES > 400 F.	(18) PENNWALT, "KYNAR ENG. PLASTICS" K68A–20M–6/69, P. 8	–	
33	LIP SEALS OPERATED SATISFACTORILY IN H–1 ENGINE AFTER 9 YEARS STORAGE. CREEP AT 160 F AND 1,000 PSI LEVELS OFF AT 12 HOURS.	AVOID CONTINUOUS EXPOSURE TO TEMPERATURES > 350 F.	(19) IBID (20) R/D REPORT MPR–8–175–519, DATED 1–18–69	(m)	
34	THREE YEAR OUTDOOR EXPOSURE SHOWS ACCEPTABLE PROPERTIES. USED FOR 11 YEARS ON RA5C VIGILANTE AIRCRAFT WITH NO AGE FAILURES.	AVOID CONTINUOUS EXPOSURE TO TEMPERATURES > 450 F.	(23) WADC TECHNICAL REPORT TR 59–664, PART II, DATED OCTOBER 1962 (24) BLOOMINGDALE RUBBER CO., DATA	(m)	
35	SEALS WITH SATISFACTION AFTER STORAGE FOR 5 YEARS AT ROOM TEMPERATURE.	AVOID EXPOSURE TO TEMPERATURES > 200 F – POOR MECHANICAL PROPERTIES.		(s)	
36	NR TESTS ON ADHESIVES INCLUDE APPLICABLE PRIMERS (MSS 003, 005, 027)	(NOT APPLICABLE)		(s)	
37	PRIMARILY ELECTRONIC CONTROL APPLICATIONS.	SUBJECT TO ABRASIVE DAMAGE.	(42) McGRAW–HILL – HDBK OF MTLS. AND PROCESSES FOR ELECTRONICS	(o, u)	
38	TESTS ON 3–1/2 YEAR OLD PART MEETS REQUIREMENTS. MOISTURE RESISTANCE 0.48% AT 60 PERCENT AND 85 F.	DO NOT USE ON PARTS REQUIRING ABRASION RESISTANCE.	(26) NR–TR322365	(q)	
39	CHEMICALLY SIMILAR TO MYLAR. EXCELLENT OXIDATION, HUMIDITY AND WATER RESISTANCE.	AVOID CONTINUOUS SUNLIGHT ND TEMPERATURES > 350 F.	(15) IBID, P. 998 (16) IBID, P. 333	(t)	
40	PROTECTED WITH PHENOLIC VARNISH – HEAT RESISTANT UP TO 300 F $\geq$10 YEAR AGING AT ROOM TEMPERATURE.	BRITTLE – UNSATISFACTORY EDGE IMPACT OR EXCESSIVE FLEXING.	(44) HDBK. OF PLASTICS – D. Van NOSTRAND CO. 1943, PP. 342–343	(m, t, v)	
41	TEN YEAR AGING AT 70 F HAD VERY LITTLE EFFECT ON PROPERTIES.	AVOID CONTINUOUS SUNLIGHT AND TEMPERATURES > 350 F.	(25) DuPONT LETTER TO NR DATED 5–14–71	(r)	
42	DATA AT 150 F AND 110 PSIG O_2 EXTRAPOLATED TO LIFE > 10 YEARS.	SUBJECT TO EXCESSIVE WEIGHT LOSS IN HARD VACUUM.	(46) HUMBLE OIL CO. DATA SHEET DG–3C, DATED 4–17–70	(e, f, l, r)	
43	QUALIFICATION UNITS OPERATED 3,000 HOURS STORED FOR 1–1/2 YEARS AND OPERATED CONTINUOUSLY FOR 1.6 YEARS.	AVOID PROLONGED SERVICE AT TEMPERATURES > 300 F.	(29) "MOTOR BEARING LUBRICANT TEST PROGRAM FOR NR APOLLO," PART XVIII, DATED 7–24–70, REPORT NO. 2187–70–380	(r, s)	
44	CHEMICALLY INERT – PASSES OXYGEN IMPACT TESTS.	LOW TEMPERATURE LUBRICITY DEGRADES.	(45) DuPONT TECH. BULS. L–5 AND G–5	(f, l, r)	
45	BALL BEARINGS RAN 11 YEARS AT 2 X 10–7 10 RR AT 130 TO 220 F. OPEN BEARINGS OPERATED AT 3,600 RPM AT 284 F FOR 1 YEAR – NO FAILURE	AVOID EXTREME PRESSURE LOAD APPLICATIONS.	(27) GENERAL ELECTRIC, SILICON PRODUCTS DEPARTMENT, TECHNICAL DATA 300K, S–21, pp. 4–9	(c, f, r, s)	
46	INERT LUBRICATING INGREDIENTS WITH INORGANIC OR ORGANIC BINDERS. ORGANIC BINDERS ARE SILICONE, EPOXY OR PHENOLIC (SEE MSS 001, 003, 004 FOR AGE LIFE DATA). INORGANIC BINDERS ARE INERT.	AVOID HIGH HUMIDITY AND MOISTURE ENVIRONMENTS AND HIGH PRESSURE APPLICATIONS.	(28) R. V. SHELTON, "FUNDAMENTALS OF FRICTION AND WEAR, BEARINGS AND THEIR LUBRICATION AND SPACECRAFT LUBRICANT," NR/SD MATERIALS COURSE LECTURE, PP. 2.27 TO 2.27.7	(n, s)	

(continued on next page)

Table 3-4 Room temperature (Earth environment) storage life of common materials used in spacecraft and launch vehicles (Sheet 4)

ITEM	COMMON NAME	TRADE NAME	ACRONYMS	CHEMICAL NAME	CLASS	FAMILY	SPECIFIC WEIGHT	AGE LIFE (YRS.)		PRINCIPAL USES
								DOCU-MENTED	MIN. PRE-DICTED	
47	ABLATOR	AVCOAT 5026-39	–	–	COMPOSITE ABLATOR	FILLED EPOXY-PHENOLIC	–	6	15	USED ON APOLLO CM HEAT SHIELD
48	CORK	INSULCORK 2755	–	–	THERMAL INSULATOR	–	–	10	20	THERMAL INSULATON AND ABLATOR FOR SPACECRAFT AND BOOSTER APPLICA-TIONS
49	CONNECTORS AND TERMINAL BOARDS	–	–	–	–	–	–	10	15	SPACECRAFT ELECTRICAL WIRING TERMI-NATIONS
50	THREAD SEALANT	LOCTITE	–	(PROPRIETARY)	SEALING AND LOCKING COMPOUND	CYANO-ACRYLATE	–	10	20	FOR SEALING AND RETAINING THREADED FASTENERS AND FITTINGS
51	PERMANENT MAGNETS	–	–	BARIUM FERRITE	MAGNET	FERRITE	–	10	–	GENERAL INSTRUMENTATION AND ELEC-TRICAL APPLICATIONS
52	PRESSURE SENSITIVE ADHESIVE TAPES	–	–		–	–	–	10	10	WIRE WRAPPING, NAMEPLATES, PROTEC-TION OF SHARP CORNERS
53	ALUMINUM ALLOYS	–	–	–	METAL	ALUMINUM	2.62 – 2.82	10	25	SPACECRAFT STRUCTURE – 1100 SERIES USED FOR SEALS
54	HIGH CHROME, CHROME/NICKEL, NICKEL, AND NICKEL BASE ALLOYS (LOW STRESS CONDITIONS)	–	CRES (CORROSION RESISTANT STEEL)	–	METAL	FERROUS	7.5 – 8.0	10	50	PIPING, ELEVATED TEMPERATURE APPLI-CATIONS, SEVERE CORROSION ENVIRON-MENT APPLICATIONS (LOW STRESS CON-DITIONS)
55	HIGH CHROME, CHROME/NICKEL, NICKEL, AND NICKEL BASE ALLOYS (HIGH STRESS CONDITIONS)	–	CRES (CORROSION RESISTANT STEEL)	–	METAL	FERROUS	7.5 – 8.0	10	15	SAME AS ABOVE – EXCEPT USEFUL IN HIGH STRESS CONDITIONS
56	COPPER AND COPPER ALLOYS	–	–	–	METAL	COPPER	8.9 – 8.95	10	25	ELECTRICAL WIRING, CONTACTS, TERMI-NATIONS, HEAT SINKS LIMITED USE FOR SEALS
57	ELGILOY WIRE AND STRIP	ELGILOY	–	COBALT-NICKEL	METAL SPRINGS	COBALT	–	10	20	CRYOGENIC SPRING APPLICATIONS – GOOD FATIGUE RESISTANCE
58	TITANIUM AND ALLOYS	–	–	–	METAL	TITANIUM	4.43 – 4.73	10	25	SPACECRAFT STRUCTURE

Table 3-4 Room temperature (Earth environment) storage life of common materials used in spacecraft and launch vehicles (Sheet 4 continued)

ITEM	REMARKS	SERVICE ENVIRONMENT LIMITATIONS	REFERENCES	STABILITY IN SPACE	SPACE ENVIRONMENT COMMENTS
47	SAMPLES TESTED 4 YEARS AFTER MANUFACTURE WERE SATISFACTORY. SAMPLES VACUUM AND TEMPERATURE CYCLED WERE ALSO SATISFACTORY. BASIC INGREDIENTS HAVE UP TO 10 YEAR LIFE. (MSS 003, 004, 015)	(NOT APPLICABLE)	(33) AVCO REPORT NO. RP412-1 "APOLLO CM HEAT SHIELD ABLATOR SKYLAB MISSION DESIGN REQUIREMENTS VERIFICATION REPORT," 28 AUG. 1970, PP. 27, 28, 30, 39, 40	(m)	(SEE PRECEDING PAGES)
48	AMSTRONG INSULCORK 2755 ON MISSILES STORED FOR ALMOST 10 YEARS SHOW NO AGING DEGRADATION.	AVOID CONTINUOUS OPERATION AT TEMPERATURES > 200 F.	(30) LTR. FROM J. C. MITCHELL, ARMSTRONG CORK CO. TO NR, DATED 4–30–71	(e, p, t)	
49	CONNECTORS AGED 6 TO 7 YEARS PASSED REQUIREMENTS. CONNECTORS AND TERMINAL BOARDS EXPOSED TO 200 F FOR 700 HOURS PASSED REQUIREMENTS. MATERIALS HAVE UP TO 10 YEAR LIFE (MSS 001, 003, 004, 005, AND 024).	(NOT APPLICABLE). SEE MSS 001, 003, 004, 005, AND 024.	(31) NR–TR332357	(n, s)	
50	COMPOUND RETAINS MORE THAN 50 PERCENT OF TORQUE AFTER EXPOSURE TO 300 F FOR 1,000 HOURS (EXTRAPOLATES TO MORE THAN 10 YEARS AT ROOM TEMPERATURE).	AVOID CONTINUOUS EXPOSURE AT TEMPERATURES > 300 F.	(32) SPEC. MIL–S–22473, "SEALING, LOCKING, AND RETAINING COMPOUNDS"	(s)	
51	ELECTRICAL PROPERTIES NOT AFFECTED BY TEMPERATURE. STABLE AT ROOM TEMPERATURE UP TO 20 YEARS.	SUBJECT TO DEGRADATION IN OXIDIZING ENVIRONMENT.	(47) FUNK & CARROLL, "ELEC. ENGS. HDBK.," PP. 4–120 THRU 123, AND 5–2 THRU 5–4.	(q)	
52	TAPE WITH SILICONE ADHESIVE SATISFACTORY AFTER 7–1/2 YEARS INSTALLATION IN SM008. BACKING MATERIALS HAVE > 10 YEAR LIFE. (SEE MSS 008, 011, 018, AND 021).	SELECT FOR SERVICE TEMPERATURE. AVOID DIRECT LOADING	–	(s)	
53	THE 1100, 3000, 5000, AND 6000 SERIES ALLOYS HAVE UP TO 20 YEARS RESISTANCE TO WEATHERING UNDER RELATIVELY SEVERE ATMOSPHERE. THE 2000 AND 7000 SERIES MUST BE PROTECTED WHEN EXPOSED TO WEATHER, HOWEVER, IN DRY INERT STORAGE CORROSIVE DEGRADATION SHOULD NOT OCCUR.	(SEE REMARKS)	(34) METALS HDBK. V01. 1, PP. 916–925	(j, q)	
54	SURFACE FINISH AND MECHANICAL PROPERTIES SATISFACTORY AFTER 10 YEARS URBAN AND RURAL ATMOSPHERE EXPOSURE.	USE DETERMINED BY SPECIFIC DESIGN REQUIREMENTS.	(34) IBID, PP. 552–564 (35) "CORROSION RESISTANCE OF METALS AND ALLOYS" BY LAQUE AND CAPSON (36) ARMCO STEEL CORP. TECH. PUB. (37) ALLEGHANY LUDLUM STEEL CORP. TECH. PUB. ON AM 350 AND AM 355 (38) DMIC REPORT 223, JANUARY 3, 1966	(q)	
55	THE AUSTENITIC (300 SAE SERIES) STAINLESS STEELS A286, PURE NICKEL AND NICKEL BASE ALLOYS REQUIRE A SEVERE ENVIRONMENT TO CAUSE STRESS CORROSION. THE OTHER ALLOYS ARE BEING ANALYZED FOR LIFE UNDER STRESS CORROSION CONDITIONS.	USE DETERMINED BY SPECIFIC DESIGN REQUIREMENTS.	(39 MCR 12943 (34) IBID (35) IBID (36) IBID (37) IBID (38) IBID	(q)	
56	TESTS CONDUCTED UNDER SUPERVISION OF ASTM CONFIRM SUITABILITY OF COPPER AND COPPER ALLOYS FOR ATMOSPHERIC EXPOSURE.	SUBJECT TO OXIDATION WITH REDUCED RELIABILITY.	(34) IBID, P. 984	(k, q)	
57	PASSED NAVY 500 HOUR SALT SPRAY TESTS – NO CORROSION.	SUBJECT TO SURFACE DAMAGE WITH BAD HANDLING.	(48) TECH. PUBLICATION OF ELGILOY CO.	(q)	
58	RESISTS CORROSION IN FERRIC CHLORIDE, SODIUM HYPOCHLORITE AND SIMILAR SOLUTIONS.	CORRODES IN FUSED–CHLORIDE SALTS, RED–FUMING NITRIC ACID AND 10 PERCENT HCL AT 95 F.	(35) IBID (49) NASA, SP–3–25 (TR64–40) (50) SUPPLEMENT 19 TO NR, SD71–267	(q)	

nuclear stages. It was concluded that of some 40 organics surveyed, only a few set the limit in nuclear engine environment, such as Teflon. It was also determined that selected materials substitutions or minor design changes would make virtually all S-IV and S-II launch-vehicle components examined withstand dosages in excess of those predicted.

Other possible exceptions are the radiation fields of some planets, including Earth, but particularly of Jupiter. Possible effects in these environments have no bearing upon Earth-launched booster engines but may have to be considered for orbital maneuvering engines, deep-space attitude-control thrusters, planetary orbital and landing systems, and return-to-Earth propulsion systems.

In Table 3-4 in the "AGE LIFE (YRS)" column, the term "DOCUMENTED" refers to life spans based upon concrete evidence, and "MIN. PREDICTED" denotes a reasonably reliable minimum life estimate, based on other observations, expert extrapolations, and comparable data for similar materials. In a few instances, no entry was made under "PREDICTED," to indicate that a reasonably reliable statement could not yet be made.

In the "REMARKS" column, the terms "AGING" or "ACCELERATED AGING" (at a stated temperature) are used. These terms reflect the terminology used in the references upon which these entries are based. It should be noted that reservations do exist about the validity of accelerated aging of nonmetallics, if it is claimed that these tests are representative of, or equivalent to, real-time aging. However, when treated as test data without claimed relation to real-time effects, the information is valid.

The table lists a total of 58 materials. Of these, seven are metals or metal alloys. All others are nonmetallic materials for which the information is especially valuable.

Apart from the table, the applicable literature, as well as consultations with cognizant workers in the field, reflects a rather uniform opinion that metals are of considerably less concern than are nonmetallic materials in regard to life capabilities. Our subsequent discussions, therefore, will concentrate on nonmetallic materials.

For 53 of the 58 materials listed, documented life spans of 10 yr are shown. For 47 of them, predicted lifetimes of 15 yr or more are indicated.

In summary, the predicted age-life of almost all materials studied exceeded 15 yr. The relatively large selection of materials thus available for most foreseeable needs should permit and justify the avoidance (i.e., the "design-out") of those few that are known to be age-limited and the substitution of another (better) material. Our observation supports the notion that, at the present state of the art, and from an age-life point of view, reliable knowledge of a material's composition, properties, and thermal history is a more dependable guide for long-life component design and verification than are accelerated life tests, which are often questionable. An additional thought is that new materials are continually being developed. Within 15 yr from the date of the tabulation, if materials should be needed with mandatory lifetimes of more than 20 yr, it appears reasonably certain that such materials will have been developed—or that at least many of the existing ones will have been proven-out beyond the 20-yr requirement.

The foregoing reasoning is further supported by the observations made when several pieces of hardware, which had spent 2 1/2 yr in space, were examined. This singular event took place in 1969. During that year, on 20 November 1969, Apollo 12 astronauts Pete Conrad and Al Bean retrieved and returned to Earth several structural members and a TV camera from the unmanned Surveyor III spacecraft that had soft-landed on the moon on 20 April 1967, after having been launched on top of an Atlas-Centaur rocket on 17 April 1967. Astronaut Conrad describes this unique event in these words:

> The Surveyor was covered with a coating of fine dust, and it looked tan or even brown in the lunar light, instead of the glistening white that it was when it left Earth more than two years earlier. It was decided later that the dust was kicked up by our descent onto the surface, even through we were 600 ft away.
>
> We cut samples of the aluminum tubing, which seemed more brittle than the same material on Earth, and some electrical cables. Their insulation seemed to have gotten dry, hard, and brittle. We managed to break off a piece of glass, and we unbolted the Surveyor TV camera. Then Al suggested that we cut off and take back the sampling scoop, and so we added that to the collection.[2]

The thorough examination of these parts has been documented in a 1972 NASA/Jet Propulsion Laboratory report.[3] According to this report, the astronauts removed the following material from Surveyor III:

The complete television camera with its associated optical and mechanical components, electronics, pieces of cabling, and support struts;

The scoop from the soil mechanics surface sampler;

A 19.7-cm section of unpainted aluminum tube from the strut supporting the radar altimeter and Doppler velocity sensor;

A 10-cm section of aluminum tube from one of the camera support struts. This tube was coated with inorganic white paint;

About 13-cm of television cable, with its wrappings of aluminized plastic film.

The conclusions drawn from the investigation of these parts are reflected in the following excerpts from the report:

The hardware taken included samples representative of many spacecraft engineering subsystems, with a wide variety of electrical and electronic components, optics, functional mechanisms, lubricants, and temperature control devices and coatings.

The results of the engineering investigations were essentially "nonspectacular"; the primary value lies in the fact that no failures or serious adverse environment effects on the hardware were uncovered that, to some degree, had not been anticipated. The absence of detected major effects and the resulting implications for future space vehicles are significant. However, the absence of effects should not be construed to indicate that the problems associated with material and component selections, test, design, assembly, and systems test can be ignored.

Almost all exposed surfaces on the camera were partially covered with a fine layer of lunar dust. The dust distribution indicates that the fines were disturbed and implanted upon the spacecraft primarily by the initial Surveyor landing and by the approach and landing of the Apollo 12 Lunar Module (LM); however, the presence of dust, even in very small quantities, can have a significant effect on temperature control and optical performance of hardware on the lunar surface.

The report conveys a feeling for the importance placed on distinguishing between effects caused by the true space environment and those caused by mission-related events, such as launch, transit, and landing, by handling of the parts during return to Earth, and by the effects incurred after return, during subsequent protracted examinations.

On 12 January 1990, the crew of the Shuttle Columbia retrieved from orbit a satellite, the 11-ton Long Duration Exposure Facility (LDEF) that had spent over 6 yr in space. At the time of this writing, preliminary conclusions drawn after its return to Earth, based on the satellite's exterior appearance, were that it appeared to be somewhat weathered and discolored. "There are lots of small micrometeoroid craters and some tearing of foils covering some of the experiments," a project member observed. "But that's why we sent it up there, to find out what happens to materials after a long time in the space environment. We are delighted with the condition."

The thorough and complete evaluation of all elements, including those below the satellite's surface, will take considerable time. However, when completed, the results will be a major contribution to the planning and design of long-duration spacecraft and the space station. Here again, the evaluations must ensure the identification of effects possibly caused during the satellite's launch preparations, its launch, retrieval and landing, and its handling on Earth.

We should not leave unmentioned the many successful unmanned deep-space probes. Although we will never be able to examine their various systems, their spectacular achievements are indirect proof that all components were in working condition after spending many years in the space environment. Especially impressive is the journey of the Voyager spacecraft, which culminated in its flyby of Neptune and its moons on 24/25 August 1989, 12 yr after launch. Its travel also included visits to Jupiter and Saturn and to a number of their moons.

In a briefing to visitors at the Jet Propulsion Laboratory in Pasadena on the day of Voyager's Neptune flyby on 25 August 1989, a staff member stated: "All systems worked flawlessly. No radiation effects were noted; if there were, they were below detection levels. Our instrumentation registered dust particle impacts in the ring planes which scared us—but there was no damage, no effects. Earlier, at Jupiter, we had some radiation damage, but we were able to restore the affected systems by various means, and regain full experimental capability."

A "Family Tree" of Nonmetallic Materials

The nomenclature for nonmetallic materials may sometimes become confusing to the designer who is not a materials specialist. We are, therefore, including a "family tree" (Figure 3-15), reflecting the terminologies and relationships of polymeric materials (some of which are used in Table 3-4). The names of major chemical companies producing one or more of them are also included.

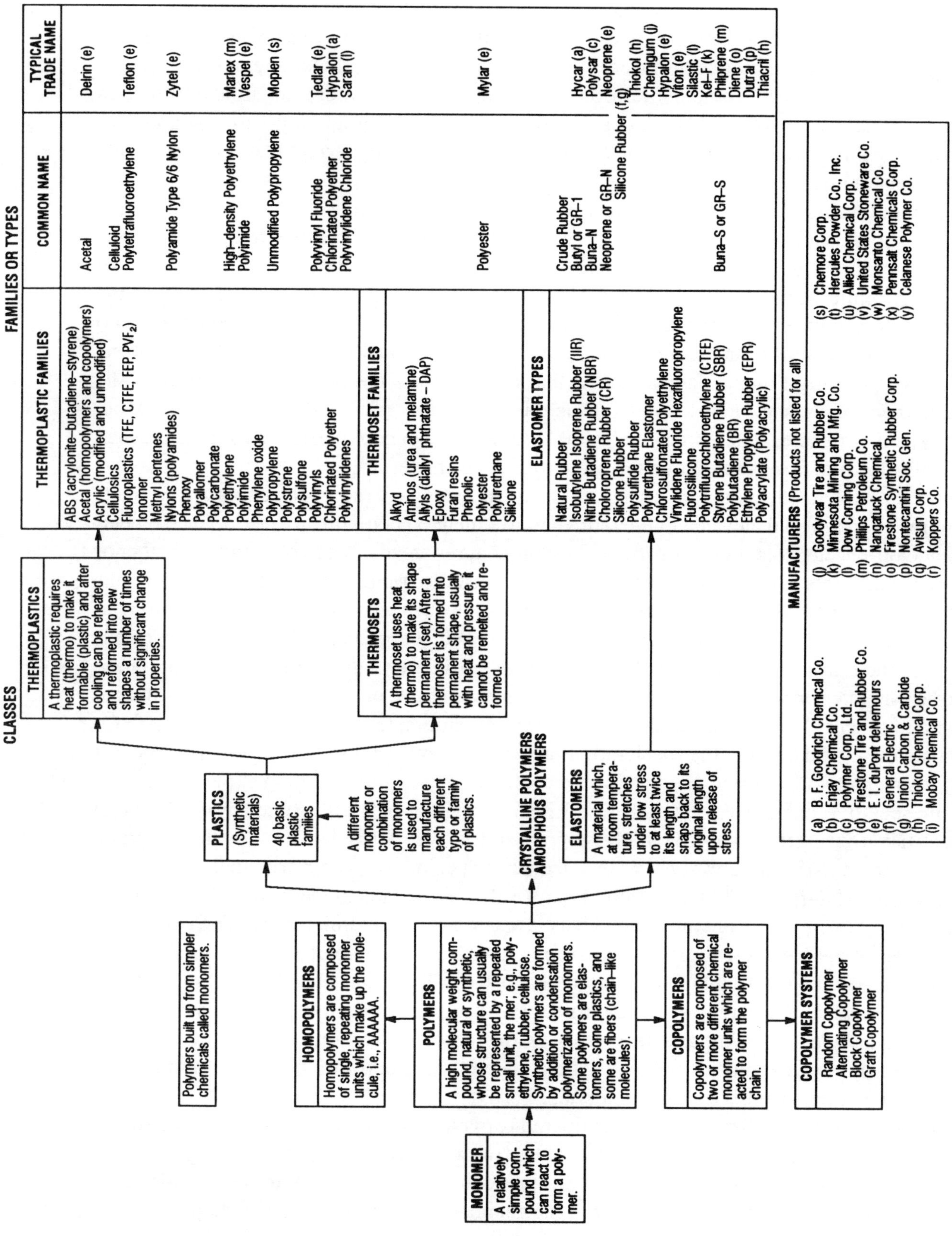

Figure 3-15. Polymeric Materials—Terminologies and Relationships

Accelerated Aging

It is a safe assumption that Table 3-4 does not cover all of the materials that are available or that are being used, or that will be used in current or future rocket engines and space vehicles. For some of those materials not listed, comparable information may exist somewhere, and little information may be documented for others. The latter situation has often led to attempts to "simulate" or compress time through Accelerated Aging of components and their materials, because of schedule and cost constraints.

Accelerated Aging may be defined as the process of obtaining effects—through application of increased load(s), such as temperature (usually elevated), pressure/vacuum, stress, voltage, etc., in a relatively short time (weeks)—which are completely representative of those resulting from much longer time spans (years) in the design environment. These effects encompass the rheological (crystalline) and morphological (chemical and/or gravimetric, i.e., componential) changes in a material as a function of time, in a stated operational environment (temperature, pressure, surrounding fluids). It is assumed that aging affects the physical properties of the materials themselves (e.g., tensile strength, hardness, specific gravity, elasticity, etc.). More generally, aging denotes deleterious, undesirable (albeit unavoidable) effects, as distinguished from controlled processes to improve the properties of the materials.

A typical Accelerated Aging test is set up to study, measure, and record the cumulative effects of combined vacuum and elevated temperature environment on component performance repeatability and on selected materials properties. Test procedures are also set to determine whether or not there are differences of behavior between stressed (installed) and unstressed (shelf) materials. The test includes finding whether there are differences between a component which is left in an environment without interruption (such as in the case of a deep-space vehicle) and one which has returned to the Earth environment cyclically (such as the shuttle). During the test, information should be generated on the minimum (optimum) compressed experiment time required to simulate space environment effects on selected materials at the maximum temperature considered safe for these materials. And because the acceleration factor is generally not known in advance, the test should also attempt to ascertain the factor of acceleration present in the aging experiment.

A representative, exhaustive Materials Age-Life test program can be illustrated by the thorough approach taken during the Long-Life Study program at Rockwell Space Division. In this program, two spare Saturn S-II 8-in. prevalves were selected as specimens. Their principal nonmetallic materials were Mylar, KEL-F, and Teflon. For all valve piece parts made from these materials, uninstalled spares were also obtained, as well as raw samples of these materials. The two valves and the material samples were placed in two ovens kept at a temperature of 340°F. In turn, the two ovens were placed in two vacuum chambers kept at 3×10^{-7} Torr. One chamber was run without interruption for 42 days, whereupon valve and materials samples were removed for valve actuation and performance tests (timing and leak rates, with due consideration of the actuating fluid temperature) and for laboratory inspections of the materials. To perform the same examinations, we opened the other chamber in 7-day intervals.

This procedure of periodic vacuum and temperature cycling was aimed at evaluating the effects of valve exposure to a cyclic operational environment of shuttle-type vehicles on timing (response and travel times) and on leak rates. In using this procedure, we also sought to better understand the trends in Materials Aging as a function of (accelerated) time (intermediate data points).

Thermal Analyses

The vacuum chamber Accelerated Aging tests (described earlier) predominantly emphasized assurance of the validity of such tests for installed materials, using assembled valves. Laboratory tests of the material samples for their physical properties as a function of exposure time were limited because the materials had been exposed to only one constant temperature level.

For a more complete exploration of materials properties, specialized thermal analyses are recommended which can be performed by test units, such as the DuPont 900 and 990. These units can perform the following tests:

Differential Scanning Calorimetry (DSC)

This test measures changes in specific heat as a function of temperature; the results are a measure of the change of heat capacity (Cp) of the material. These data, in turn, permit the determination of the fusion heat and of other phase changes.

Differential Thermal Analysis (DTA)

The function of this test is to measure the thermal behavior of a material caused by physical and chemical changes during heating and cooling. The test yields these results:

The temperatures at which the material and its constituents undergo various changes, such as boiling, melting, thermal decomposition, and crystal phase changes

Heat absorption (endotherm) and heat rejection (exotherm) of the material as a function of temperature

Thermal Mechanical Analysis (TMA)

As a function of temperature, this test measures the changes in a material, specifically as they may affect the following parameters:

The coefficient of thermal expansion under compressive load
Compressibility
Tensile strength
Volumetric coefficient of expansion

Prospects of Accelerated Aging

Studies of the literature and the results of laboratory tests which were performed indicate that reliable accelerated aging of mechanical components, containing nonmetallic materials, or the accelerated aging of material samples, is not practical or reliable, if, in fact, it is even feasible. Several reasons for this conclusion follow:

Up to temperature levels characteristic for a given material, most materials do not degrade over long periods of time. But by exposing them to higher temperatures in an attempt to accelerate their "aging" (compressed time), effects are produced which are not representative of those which would occur at, or below, the maximum design level. *Increased temperature acts as a modifier, rather than as an accelerator.*

The various nonmetallic materials of interest to the designer of long-life mechanical and electromechanical space- and launch-vehicle components react differently—often significantly so—to increased temperatures. If one temperature is chosen as the "load," to cause an acceleration of the life testing of a component containing several nonmetallic materials, this temperature could not be, under any circumstances, the right one for all of the materials contained therein. This problem may be expected to become more serious, as nonmetallic materials are being developed with significantly increased maximum allowable operating temperatures.

Even if a component contains only one type of material, reliable accelerated life testing is not considered practical, for the same reasons stated before. Thermal analysis data have shown that most, if not all, polymeric (nonmetallic) materials undergo changes at discreet temperatures, often in relatively short timespans. These changes do not occur if the materials are kept below these temperatures, even over very long periods of time (15 yr or more). The application of temperatures higher than the maximum design value introduces changes unrepresentative of "real-time aging."

For similar reasons, materials tests and thermal analyses, especially short term, cannot provide reliable life prediction for the various nonmetallic materials where they are used as seals, bearings, or other pieceparts in mechanical and electromechanical components. Such tests do yield, and in fact have yielded, the data needed to describe the properties of the materials and to aid with their proper selection. But these tests will not predict their real-time behavior following machining, forming, and installation (stress) or their longevity when used at the dynamic interfaces of operational components.

We may conclude, then, that there are numerous nonmetallic materials now available which retain their properties for 10 to 20 yr in many environments encountered by space- and launch-vehicle components. Since the life requirements stated for the mechanical and electromechanical components of space and launch vehicles of the foreseeable future generally fall into a span of 10 to 15 yr, what is really needed to demonstrate the lifespan capabilities of the components? Based on the foregoing observations, the conclusions are that instead of performing accelerated materials/component life tests of questionable outcome, the emphasis should be shifted toward rigidly controlled materials manufacturing processes, assurance of the use of the correct material in the proper application, and greater awareness of their limitations and/or needed servicing procedures.

Unlike the accelerated tests with questionable outcome which we have described, accelerated cycling *does* lend itself to reliable acceleration in a carefully designed test program. After all, the primary concern of the designer and the user of mechanical and electromechanical components is that the components perform reliably under all specified operational environmental conditions and over specified lifetime spans. What happens during the component's lifetime to the materials of which it is composed is of secondary concern, as long as these effects do not cause component performance to deteriorate abnormally, i.e., to fall out of specified limits before achieving design life and/or operational cycles.

One of the shortcomings of much of the literature devoted to materials properties is that this simple fact is not sufficiently considered and that these treatises do not, and often probably cannot, supply any information which would permit the designer to estimate, predict, or influence component operational life at the time of materials selection and component design. A major reason for the limitations of much of the published data is the wide range of conditions under which the materials have to perform—fabricated, treated, stored, machined, and formed before installation—and stressed, exposed, operated, and serviced following installation.

It is apparent that a reliable answer can be obtained only through endurance testing of the component in its carefully duplicated, operational environment, using the dynamic Interface/Weak-Link Principles, including systematic wear analyses. The expectation is that tests performed using these principles with a complete component, or where possible, with individual interfaces, will result in more economical and more reliable development tests, over a shorter timespan.

Some Words of Caution

IN Chapter 2, we introduced the two principal "enemies" of component longevity: Wear and Materials Aging. Means to reduce or eliminate these two effects have been a major subject of our discussions. However, we also listed in Chapter 2 a selection of additional mechanisms that may affect component life and which cannot be prevented (at least, not entirely) by the dynamic interface principle of identifying and correcting weak links of components.

In the following, we are listing an additional number of potential problems which are not preventable through Long-Life Technology alone. Many areas exist in which the most sophisticated analyses will not be able to forestall problems unless past experience, lessons learned, mutual consideration between disciplines, and common sense are fully applied during a component's design, development, and operational phases. Typical problem areas are summarized below, in no specific order. Some of these have been touched upon earlier. Problems have arisen repeatedly in all of these areas in the past and have affected, sometimes seriously so, a component's performance, reliability, and the ability to qualify, as well as its utility and maintainability.

Weight and Size

The purpose of all rocket-propelled vehicles is to accelerate their payloads to the speeds required for their intended trajectories, including those required for placing a vehicle into orbit. The weight of a rocket vehicle consists of gross (tanked) vehicle weight and payload weight. The vehicle designer attempts to achieve the best possible ratio between the two, in favor of the payload. This effort includes the minimization of the cost per pound of a given weight launched over a given distance or placed into orbit. Unfortunately, the hard facts of life are not very supportive of these endeavors and sometimes lead to intense pressure on the designers to cut vehicle weight. The prospects of achieving flexibility for most types of hardware appear dim without adversely affecting a component's integrity; the best to hope for in light of the increased mission requirements is that the line can be held. The same applies to shape, size, and envelope.

Sophistication

Every effort to simplify space hardware without loss of function is strongly urged in order to reduce both failure potentials and the cost of fabrication, development, and maintenance. It should be realized, however, that the component designer must not arbitrarily omit or alter a feature that has been specified by the system.

Material Selection

A substantial number of proven materials are now available that are considered reliable candidates for future space hardware. The designer's thorough familiarity with these materials is mandatory in order to avoid problems. Substitutions should not be attempted without good reasons and without expert consultation. (See also Table 3-4.)

Materials Processing and Fabrication

Apart from the normal mission environment, the materials constituting the mechanical components of a space vehicle may be affected by abnormal and erratic external influences. For instance, various manufacturing processes (adhesive bonding, protective metal coatings and treatments, fluxes, solvents), fabrication techniques (welding methods, brazing, soldering), and production testing methods (leak, X-ray, N-ray) may interact with component constituents, introducing contaminants and otherwise affecting the rate of material degradation. These processes, if uncontrolled, often have a strong influence on component longevity or act as a catalyst in modifying the effects of normal aging. Consequently, such interactions must be considered in the development of long-life components.

Component Timing

The many concepts available to control the kinematics of components containing moving parts are well-understood and should not cause any deficiencies, if properly applied.

Maintainability

It may happen that certain components (capable of meeting a requirement of 100 flights, for instance) cannot be developed economically. In this case, routine replacement at scheduled intervals may be less costly overall, and designs that permit such maintenance in a routine fashion may be necessary. Even components that are designed and developed to last for 100 missions may nevertheless require occasional replacement or service. It is evident that an optimum method, acceptable to all, must be developed.

Ease of Checkout

Because ease of checkout between flights is a basic requirement of the reusability concept, the suitability of the hardware (existing or to be designed) for easy checkout will be a critical design aspect.

Accessibility

The practicality of both component maintenance and component checkout will depend to a considerable degree on accessibility. The task of the vehicle systems designer will be easier if the component design involves maintenance concepts from the outset: for example, the design would allow for the replacement of a subassembly, such as a pilot mechanism in a primary valve, instead of routine replacement of the entire valve—and by using simplified procedures, especially in crowded quarters.

Mismatch

The strong emphasis upon thoughtful planning, to avoid such problems as mismatches, reversed hookups, crossed actuation lines, and reversed electrical connectors during initial assembly and component replacements, might appear to be elementary—if it were not for the embarrassing events that continue to occur. Prevention of these mistakes requires careful attention, which should not be limited to the initial design phase; planning should include, for example, the alternate use of existing components in new systems. Effective preventive measures must also be applied to various areas of concern, such as clear labels, orientation arrows, dissimilar connectors, and reference designators.

Communications

Poor communications between customer/user offices and contractors, associated contractors, and subcontractors have often caused serious problems, resulting in costly redesign (to make fit) of mechanical components and other hardware. Good communications are particularly important when common use of hardware by several contractors is planned. A rigid, complete, and reliably implemented interface control system is therefore mandatory. During the three-stage Saturn V booster program, the NASA Marshall Space Flight Center instituted, most successfully, a system of interface control drawings which showed all the necessary details and dimensions of the elements constituting an interface between the products of different contractors, all on one drawing. Typical interfaces are as follows:

Engine to stage structure
Engine to propellant (inlet) lines
Structure Stage A to structure Stage B
Vehicle to ground support equipment
Vehicle to ground servicing equipment

No dimensions or other specifications were allowed to be changed without the approval of the NASA Vehicle Project Office.

Late Changes

Realistically, though, it must be expected that some late specification changes will occur—near to, or at the time of, component design release. Because built-in flexibility must be limited for reasons of overall economy, the shortest possible line of communication between user and designer is essential.

Mishandling

The problem of mishandling vital and sensitive components is usually the result of ignorance, carelessness, or indifference. Sufficient examples exist in earlier space programs to illustrate the need for careful selection, qualification, and continuity of skilled personnel. Nevertheless, it will always be beneficial to design mechanical components that are as insensitive as possible to external influence. The use of additional, separate protective devices is advised wherever possible and practical.

Safety

The emphasis on space vehicle safety can be expected to be as strong into the future as it has been in the past with manned vehicles. The effects of accidents could be very serious indeed. Apart from fundamental manufacturing, assembly, and operational quality controls, the basic design of components can contribute significantly to safety through fail-safe concepts, selected redundancies within the component, and the use of proven concepts.

A special aspect of safety is the compatibility of component materials with various fluids—in particular, oxygen. The absolute exclusion of materials identified by government agencies and contractors as being incompatible is mandatory.

Qualification

Previous specifications that dictated sometimes substantial multiples of operational life to be demonstrated for qualification will no longer be realistic or economical. An early start is urged toward a practical methodology, with emphasis on operational capability rather than on "brute force" demonstration, to prevent operationally sound components from being tested to death and then condemned.

Corrosion

Because of the relatively large number of flights projected for major space programs, such as the Shuttle Orbiter, involving hundreds of temperature and pressure cycles, and operational time spans of several years, corrosion is a real and potential danger. However, this danger should be controllable through meticulous implementation of proven safeguards, as well as by the proper selection and treatment of materials.

Stress Corrosion

A considerable amount of experience has been accumulated during the Saturn V program. Painstaking application of past recommendations should minimize potential problems.

Fracture Mechanics

This phenomenon has represented a real risk for components exposed to cryogenics, such as feed and vent lines. However, within the last 10 yr or so, great strides have been made in defining the mechanisms of crack propagation, crack detection, and the development of fracture control plans, to the extent that most new space vehicle contracts require the submittal of fracture control plans and the performance of fracture mechanics analyses in conjunction with normal stress and dynamics analyses. In fact, many space vehicle drawings (including mechanism and propulsion component details) include general notes for inspection to detect surface cracks exceeding specified dimensions.

Contamination

The long times between overhauls and the frequency of filling the tanks (for up to 100 operational missions, such as with the space shuttle) mean that the gradual accumulation of contaminants, which might have been insignificant in one-mission vehicles, now represents a real danger—and this danger must be dealt with.

Creep and Drift

The many temperature and pressure cycles, to which most components (especially those in cryogenic systems) will be exposed, can cause a problem, the progressive deterioration of sensitive tolerances. For instance, increased sizes (sealing diameters) of many components will affect leak rates, responses to pressure changes, timing, and actuation forces. Some of this deterioration may be controllable through readjustments; for other cases, more permanent solutions will have to be developed. Closely related are the effects from performance drifts, such as those attributable to deposits forming in sensitive sensing ports, changes in reference vacua, the motion of threaded (or similar) setting devices, and the like.

Flow Dynamics

The effects of flow-induced vibration on flex lines were dramatized on the third Saturn V launch, and the lessons learned must be applied to future space vehicle designs. The theories developed and the design criteria based thereupon have been sufficiently demonstrated in experiments in order to instill high confidence in their validity. Development of these theories should continue, to ensure that even exceptional cases will be reliably covered.

Impact Loads from Waterhammer

Waterhammer, caused by sudden stoppage of propellant flow as a result of rapid valve closing or by gas bubble formation and collapse in cryogenic propellant ducts, can be decreased through larger line diameters. However, because of weight and valve-size considerations, this solution appears undesirable. Emphasis, therefore, should be directed toward exploration of improved valve timing, absorbers, special manifolding, and bubble-preventing cryogenic propellant circulation devices.

Environment

The vibration environment to which components will be exposed has been touched upon in other parts of our discussions. In addition, the designer must pay attention to the cyclic or prolonged effects of temperature, the prolonged effects of radiation, especially on nonmetallic materials, and the effects of zero-g (if any). Accurate simulation of the operational environment during development will be a difficult—but essential—task.

Interface Design Concept Tabulation

DURING the extensive Long-Life Technology Study at Rockwell International's Space Division, a total of 244 dynamic interfaces was analyzed and tabulated, representing 15 Interface Design Categories. Within each category, varying numbers of design concepts were listed. Figure 5-1 illustrates the concept tabulation format, applied to a typical interface design concept. The entries are largely self-explanatory and reflect the effort to make the tabulation complete. A few entries are specific to the quoted study, such as the reference to data books and photographs, but they were retained to exemplify the completeness which is recommended.

A table, like the sample shown in Figure 5-1, is a valuable empirical tool for the component designer and the wear-rating analyst in support of their endeavor to eliminate weak links. Appendix A to our discussions contains the entire tabulation generated during the cited study. It is unlikely that similar surveys exist elsewhere. On the other hand, it could be surmised that the tables do not include a number of dynamic interface concepts successfully used elsewhere, because they were not represented in the study. However, the tabulation should at least be a starting point and a stimulator of ideas.

By definition, any mechanical, dynamic interface, if reduced to its most elementary functional form, implies the contact between, or the exertion of pressure by, one mechanical element on another. For the component designer, however, having all interface wear analysis findings, lumped together into a single listing, would not be very useful. Thus the need to subdivide into categories becomes apparent, and the subdivision was done with the tabulation. The resultant subdivisions were primarily based on logical interface functions. In a few instances, the subdivision had to reflect those interfaces that primarily provide a sealing function and are separated from those with a load-carrying or guidance function. In turn, these primary functional divisions are subdivided into 6 sealing/seating categories and 9 bearing/guidance categories; the 15 categories are as follows:

Category No.	Interface Category Description	Design Concepts Represented
1	Reciprocating shaft seals — radial sealing	7
2	Reciprocating piston or poppet seals — radial sealing	12
3	Rotating shaft seals — radial or face sealing	4
4	Rotating main closure seals — spherical or face sealing	4
5	Reciprocating poppet, ball, or piston seats (plastic) — axial seating	11

Category No.	Interface Category Description	Design Concepts Represented
6	Reciprocating poppet or ball seats (metal-to-metal) — axial seating	12
7	Reciprocating shaft or piston guides (plastic) — radial bearing	8
8	Reciprocating guides (metal-to-metal) — radial bearing	17
9	Rotating shaft journal bearings — radial bearing	6
10	Rotating shaft ball and roller bearings and ball screws — radial bearing	10
11	Rotating pin joint bearings — radial bearing	14
12	Axial/rotating thrust bearings — face bearing	14
13	Miscellaneous locating/support guides; bearings	9
14	Axial stops and impact members (push-rods)	14
15	Ball locks (detents)	1

There is no importance-weighting intended in the listing sequence of the interface categories.

In each interface category, the interfaces are tabulated in order of increasing wear rating (increasing assessed wear capability), which was assigned during the wear-rating procedures described earlier. They could have been listed in order of increasing size, cycle life, sealing pressure or allowable leakage for seals, load levels for bearings, or any of several other major parameters. The latter are often interrelated, whereas, numerical wear rating is independent of design and functional parameters. For all interface categories, the data for each interface are tabulated under four main headings:

References to Components Data Book
Interface Description
Interface Members Description
Using Component Description

The content of the second and third subdivisions (interface descriptions) is individually tailored to include those parameters considered to be most essential to the subsequent comparative evaluation of the interfaces in each category. This format is illustrated in Figure 5-1, a typical interface category.

The Interface Design Concept Tabulations, as presented in Appendix A, represent a compilation of the cumulative interface data derived through the wear analysis of 13 mechanical or electromechanical components. The wear-analyzed components, by component-type, reveal the following:

Five cryogenic propellant shut-off valves
Three pneumatic system solenoid valves
Three pneumatic actuation system or propellant tank pressure regulators
One propellant tank vent and relief valve
One propellant directional control vent valve

		DESIGN CONCEPT NUMBER	10-1
COMPONENT DATA BOOK		CONTROL NUMBER(S)	S-II/90-1
		INTERFACE DASH NUMBER(S)	-6A & -6B
		WEAR ANALYSIS RATING (AVG.)	4.9
		PHOTOGRAPHS, PAGES	B-5, -6, -12, -13, -14, -15, & -20
INTERFACE DESCRIPTION		INTERFACE PICTORIAL DESCRIPTION	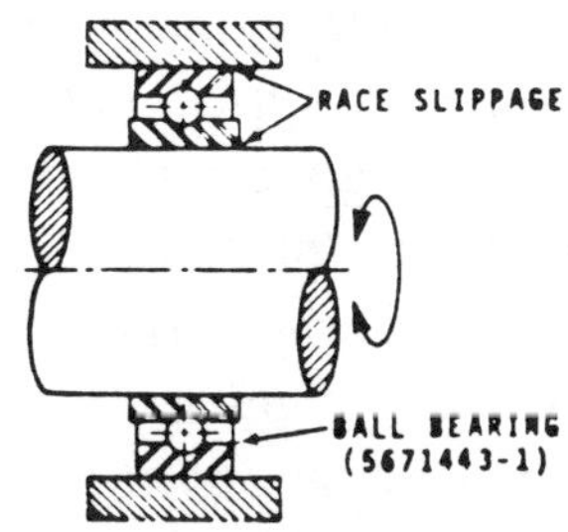
		BEARING TYPE	BALL
		OUTER RACE O.D., IN.	2.5620 ± .0005
		INNER RACE I.D., IN.	1.8125 +.0005,-.0000
		OUTER RACE DIA. CLEARANCE, IN.	.0005 - .0025
		INNER RACE DIA. CLEARANCE, IN.	.0010 - .0025
		DESIGN OPERATING MEDIA	LOX or LH_2
		DESIGN OPERATING TEMP. RANGE, F	-423 to +160
		DESIGN OPERATING LIFE, CYCLES	1000
		SPECIMEN OPERATION, CYCLES	1178
INTERFACE MEMBERS DESCRIPTION	BEARING	PART NUMBER	5671443-1
		PART NAME	BEARING - CONNECTING ROD
		MANUFACTURER P/N	RAYTEE P/N 46650B
		RACES MATERIAL	QQ-S-763 440 C CRES
		HEAT TREATMENT	R_C 58 MIN.
		SURFACE FINISH	16 rms
		SURFACE TREATMENT	PASSIVATE
	MATING MEMBERS	PART NUMBER(S)	5671434-1; 5671435-1
		PART NAME(S)	ROD; CRANK
		MATERIAL	INCONEL X750; 15 - 5 PH CRES
		HEAT TREATMENT	R_C 30 MIN.; R_{HC} 24/30
		SURFACE FINISH	125 rms
		SURFACE TREATMENT	PASSIVATE
USING COMPONENT DESCRIPTION		PROCUREMENT DOCUMENT(S) OR P/N(S)	ME284-0358-0001
		COMPONENT NAME	8-INCH PROPELLANT SHUTOFF VALVE (PREVALVE)
		MANUFACTURER	PARKER-HANNIFIN
		MANUFACTURER'S PART NUMBER	5670024
		USING VEHICLE(S) AND STAGE(S)	SATURN S-II
		USING SYSTEM(S)	LOX & LH_2 FEED

Figure 5-1. Typical Interface Design Concept Tabulation Format

A total of 244 individual interfaces was wear-analyzed, representing 143 distinct interface design concepts (which means that a number of analyzed design concepts were represented by more than one interface).

The data compilation represents a cross-sectional overview of various component manufacturers' approaches toward solving particular interface functions. The tabulation provides a sizable interface data bank from which future component designers and assessors may draw upon the experience of the suppliers of these components. The wear analysis ratings provide for rapid assessment of the relative merits of the various interface configurations. Critical reviews of these tabulations, with respect to the advantages, disadvantages, and limitations of the existing interface concepts, should further increase their value.

The tabulations compiled in Appendix A represent an important first step. Although the number of tabulated interfaces may appear to be large, they represent only a fraction of what could ultimately be envisioned for a "preferred or qualified interface design concept catalog," as part of an overall set of Long-Life Design Criteria. "Next generation" effort should include (1) addition of important design concepts not presently included, (2) inclusion of the interface design concepts, as modified, and (3) considerations for novel interface designs with dimensions, operational levels, and usage significantly different from those presently shown.

In the effort to extend the life of a component through a wear analysis, followed by modifications to eliminate weak link(s), the results may be displayed by using the interface tabulation format.

Figure 5-2 illustrates a typical modified versus unmodified (original) interface comparison. Note the significant shift in wear ratings reflected in this example. The substantial improvement achieved was the result of only two changes: a different seat retention and a different seat material. One may say that these two areas had been weak links *within* a weak link (interface). *This example shows again the danger of disqualifying an entire component for a correctable weak link.*

The quoted report[1] lists similar results for the interfaces of other components, although the step-ups were smaller, because the premodification ratings had been higher. One interface exhibited a slight reduction in rating following modification. The limited scope of this reduction signifies that the results were inconclusive and suggests that a repetition of the procedures may be required.

An initial overall appraisal of the interface tabulations reveals the following general observations:

The listing of interfaces in accordance with increasing wear rating (increasing assessed wear capability) has been demonstrated to be an effective means of presentation. No single design or performance parameter has been found to convey the relative merits of the various interfaces in each interface category, as well as the listing.

In all interface categories where comparative data and ratings are available, those interfaces containing a nonmetallic (plastic) member exhibit an overall average rating higher than the metal-to-metal contact interfaces. This higher rating is important and was not unexpected.

Within many of the interface categories, the diverse selection of mating-member material combinations, surface finishes, and surface treatments among similar functional interfaces gives the impression that there is a lack of meaningful fundamental guidelines.

			5-8	UNASSIGNED
		DESIGN CONCEPT NUMBER	5-8	UNASSIGNED
COMPONENT DATA BOOK		CONTROL NUMBER(S)	S-II/104-2A	S-II/104M1-2
COMPONENT DATA BOOK		INTERFACE DASH NUMBER(S)	-4	-4
COMPONENT DATA BOOK		WEAR ANALYSIS RATING (AVG.)	4.3	9.9+
COMPONENT DATA BOOK		PHOTOGRAPHS, PAGES	B-3 & -7	B-1 & -5
INTERFACE DESCRIP-TION		INTERFACE PICTORIAL DESCRIPTION	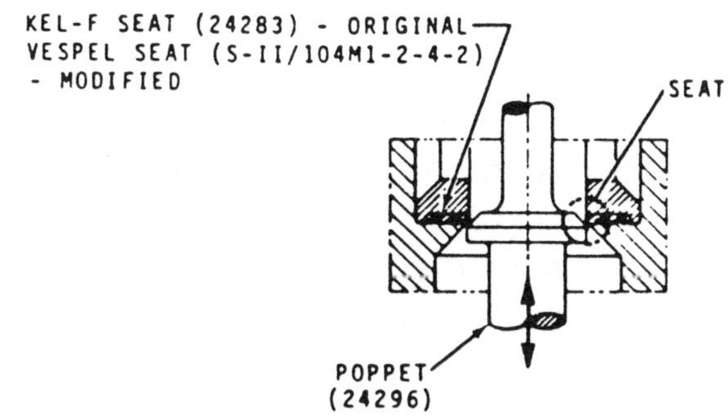	
INTERFACE DESCRIP-TION		SEAT TYPE	CONICAL	
INTERFACE DESCRIP-TION		BASIC SEALING DIA., IN.	.375	
INTERFACE DESCRIP-TION		DESIGN OPERATING MEDIA	GHe	
INTERFACE DESCRIP-TION		DESIGN OPERATING PRESSURE, PSIG	750	
INTERFACE DESCRIP-TION		DESIGN OPERATING TEMP. RANGE, F	-180 to +160	
INTERFACE DESCRIP-TION		MAXIMUM ALLOWABLE LEAKAGE	45 SCIM	
INTERFACE DESCRIP-TION		DESIGN OPERATING LIFE, CYCLES	2000	
INTERFACE DESCRIP-TION		SPECIMEN OPERATION, CYCLES	1096	10,125
INTERFACE DESCRIP-TION				
INTERFACE MEMBERS DESCRIPTION	SEAT	PART NUMBER	24283	S-II/104M1-2-4-2
INTERFACE MEMBERS DESCRIPTION	SEAT	PART NAME	SEAT, POPPET	
INTERFACE MEMBERS DESCRIPTION	SEAT	MATERIAL	KEL-F TEC GRADE B	VESPEL
INTERFACE MEMBERS DESCRIPTION	SEAT	SURFACE FINISH	32 rms	
INTERFACE MEMBERS DESCRIPTION	SEAT	RETENTION METHOD	SAGED - CLAMPED (SERRATION)	CLAMPED (SERRATION)
INTERFACE MEMBERS DESCRIPTION	SEAT			
INTERFACE MEMBERS DESCRIPTION	SEAT			
INTERFACE MEMBERS DESCRIPTION	MATING MEMBER	PART NUMBER	24296	
INTERFACE MEMBERS DESCRIPTION	MATING MEMBER	PART NAME	POPPET, UPPER	
INTERFACE MEMBERS DESCRIPTION	MATING MEMBER	MATERIAL	QQ-S-763 303 CRES	
INTERFACE MEMBERS DESCRIPTION	MATING MEMBER	SURFACE FINISH	8 rms	
INTERFACE MEMBERS DESCRIPTION	MATING MEMBER	SURFACE TREATMENT	ELECTROPOLISH	
INTERFACE MEMBERS DESCRIPTION	MATING MEMBER	OTHER SURFACE TREATMENTS	NONE	
INTERFACE MEMBERS DESCRIPTION	MATING MEMBER			
USING COMPONENT DESCRIPTION		PROCUREMENT DOCUMENT(S) OR P/N(S)	ME284-0356-0001	BASIC COMPONENT AT LEFT WITH MODIFICATIONS FOR EXTENDED OPERATIONAL LIFE.
USING COMPONENT DESCRIPTION		COMPONENT NAME	3/8-INCH NORMALLY CLOSED SOLENOID VALVE	
USING COMPONENT DESCRIPTION		MANUFACTURER	J. C. CARTER	
USING COMPONENT DESCRIPTION		MANUFACTURER'S PART NUMBER	6969-2	
USING COMPONENT DESCRIPTION		USING VEHICLE(S) AND STAGE(S)	SATURN S-II	
USING COMPONENT DESCRIPTION		USING SYSTEM(S)	VALVE ACTUATION, HELIUM INJECTION & ACCUMULATOR FILL	

Figure 5-2. Typical Modified Versus Unmodified Interface Configurations

Numerous interfaces employ the same material for both mating members—a general taboo because a high potential for adverse adhesive wear (e.g., galling) exists.

Several interfaces exhibit wear conditions that suggest a poor or haphazard selection of materials for one or both interfacing members.

A limited number of interfaces simply represent unsophisticated design—the unimaginative approach to achieving a particular interface function.

The prevailing impression emerging from the review of the tabulations concerns the apparent lack of any guidelines to applied interface design. This impression does not infer criticism of basic interface conceptualizations nor of their dimensional (fit) analyses. Rather, it challenges interface design implementation in terms of mating-member materials selections, surface-finish assignments, and surface treatment utilization, or lack of the same. The axiom that certain design provisions have been dictated by "experience" will not be good enough. If this basic premise is valid, then these experience factors must be translated into needed meaningful guidelines.

As a first step toward this end, the following detail interface evaluations, by category, will attempt to identify and highlight critical design features pursuant to extended-life interface implementation. The numbers preceding the interface category description are the assigned category numbers as sequentially listed in Appendix A.

1. Reciprocating Shaft Seals—Mylar and Teflon are primary polymers utilized for reciprocating shaft seals, although they are not represented here. KEL-F has also been successfully utilized in this seal application. There is no predominant material preference; with certain precautions, either of the two primary materials may be successfully used.

For Mylar seal applications, the following guidelines are advanced:

The favored mating-member materials of the past—A-286 and Inconel 718—are no longer recommended because of their inherently poor surface-machining characteristics. Grinding tears ("fish tails") are commonly encountered in finishing critical sealing surfaces. The materials, beryllium-copper and 304L CRES (if properly cryogenically preconditioned), have demonstrated improved machineability.

Chrome-plating the mating-member sealing surfaces has proved to be a successful technique in minimizing wear at these sealing interfaces.

A highly polished 4 to 8 rms (preferably 4 rms) mating-member sealing surface is recommended for extended operational life. High-frequency, as-lapped (matte) plated sealing surfaces are to be avoided.

Seal backup support surfaces shall be in close proximity to sealing surfaces, at least within two seal thicknesses, to avoid seal blow-back under pressurized operation. For Interface 1-1, the low rating of 4.5 was a result of observed blow-back indications during wear analysis.

Back-to-back piggyback seals represent the favored balanced installation, if the higher axial frictional loads can be tolerated.

The sealing lips of the seal blanks must be devoid of burrs, scratches, fractures, etc.

For Teflon seal applications, these guidelines are advanced:

In order to achieve consistent sealing, Teflon seals should be externally preloaded by means of garter springs or similar devices (see Design Concepts 1-2, 1-5, and 1-6). The inherent lack of rigidity, coupled with the cold-flow characteristics of this material, does not allow for initiation and maintenance of sealing capability without preloading devices.

In contrast to Mylar seals, Teflon seals are more tolerant of mating-member sealing surface irregularities. Surface finishes of 8 to 16 rms are common. In fact, 32-rms finishes have been successfully utilized on occasion. It is suspected that Teflon particle generation (wear) at these interfaces fills in the voids in the mating-member sealing surfaces, thereby enhancing sealing capability. However, the higher surface finishes are not recommended for extended operational life interfaces because of potentially higher wear. Also, attendant particle generation may migrate to, and adversely influence, other sensitive interfaces which are in close proximity. The influence of surface finish on wear and associated wear rating is exemplified in the comparison of Design Concepts 1-2 and 1-6.

In conjunction with our previous comments concerning interface wear, it appears that a filled Teflon seal provides wear characteristics which are superior to the basic polymer (again exemplified in the comparison of Design Concepts 1-2 and 1-6).

Used with the garter spring-loaded seals, a backup support ring has demonstrated its effectiveness toward the uniform distribution of seal loads. This configuration tends to minimize undesirable high heel- and/or toe-loading and associated localized wear. See Design Concept 1-6.

Although Design Concept 1-5 has functioned well in low-pressure sealing applications, especially in its modified form, it represents a "forced" approach toward achieving the desired interface sealing function. This seal configuration has evolved into a basic "labyrinth-type," subscribing to the addition of alternating seal and loading member plies until its sealing capability is achieved. Its lack of design sophistication does not make it particularly attractive for future component usage.

2. Reciprocating Piston or Poppet Seals — Design Concept 2-1 is similar to Design Concept 1-5 (previously described). The former comments apply to this interface design concept as well. In addition, the interface is used in a significantly higher pressure application, further reducing its effectiveness.

The various gross-leakage-control sealing interfaces (Design Concepts 2-2, 2-3, 2-4, 2-7, 2-8, 2-10, 2-11, and 2-12) represent a major segment of this category. Because there are no stringent performance (leakage) controls imposed on these interfaces, they must be evaluated primarily on their susceptibility to high wear.

Design Concept 2-2 illustrates a high-wear interface, as was evident upon examination of the seal in the wear analysis specimen (an endurance-tested component). In this unit, the interfacing leg of the seal was almost worn through. The abnormally high wear was attributed to utilization of a highly loaded KEL-F seal interfacing with a relatively rough 32-rms surface finish on the hard anodized mating member. Unlike Teflon, the generated wear particles of KEL-F do not usually fill the irregularities in the mating surface; hence, there is no comparable easing of subsequent high-abrasive wear conditions nor enhancement of sealing capability.

Using a filled Teflon seal and mating it with a polished 8-rms surface should substantially reduce the high-wear conditions.

Another potentially high-wear sealing interface is illustrated by Design Concept 2-8. This seal installation has a garter spring preloaded, thin Teflon lipseal, which is considered to be subject to wear-out with limited operational cycling. Because it is employed in a limited operational override provision of the component, it should provide satisfactory longevity in the application, but it is still not a promising concept for extended-to-long-cycle life applications.

Design Concept 2-4 illustrates the single, metal-to-metal contact interface in this category. The high wear (caused by smearing and metal transfer from the aluminum bronze piston rings), observed in the wear analysis specimen, is primarily attributed to interfacing the piston rings with a high-frequency, as-lapped, chrome-plated (matte) mating-member sealing surface. The high-wear conditions could be significantly reduced by imposing a high-polish requirement on the chrome-plated interfacing surface. A change in material to one of greater wear resistance, such as beryllium copper, should also tend to improve the sealing capability of the piston rings.

The remaining gross leakage control sealing interfaces (Design Concepts 2-3, 2-7, 2-10, 2-11, and 2-12) are all similar in design, differing only in the number of piston rings and the seal material. Each seal unit employs one, two, or three plastic (TFE Teflon or Fluorogold, a filled Teflon) piston rings, with a backup wafer spring to maintain contact between the rings and the mating sealing surfaces. From a wear standpoint, there is no significance, readily apparent, to the number of piston rings nor to their material. Further, there is no consistency in the assignment of surface-finish values to the piston ring nor to the mating-member sealing surfaces; therefore the following guidelines are deemed appropriate for future seal designs. The number of piston rings utilized in a particular seal unit will be left to the discretion of the designer or as dictated by specific performance requirements. Toward the goal of minimizing wear, the filled Teflons are recommended in preference to the basic polymer. A 16-rms finish is suggested for piston-ring sealing surfaces and an 8 to 16-rms finish for mating-member sealing surfaces.

The three other interfaces in the category (Design Concepts 2-5, 2-6, and 2-9) provide for tighter leakage control (specific leakage requirements).

Design Concepts 2-5 and 2-9 illustrate Mylar lipseals similar to those implemented in Interface

Category Number 1. The comments concerning the seal installations of that category are likewise applicable to the seal interfaces of this category.

Design Concept 2-6 illustrates utilization of a Teflon seal similar to those of Interface Category Number 1. Again, the comments concerning the comparable seal installations of the preceding category are equally valid to this seal interface.

3. Rotating Shaft Seals — This interface category is represented by three design concepts, employing plastic seals (Mylar and Teflon), and one design concept, using a metal "U-cup" seal. As expected, each of the seal interfaces employing a nonmetallic element reflects significantly higher wear ratings than does the metal-to-metal contact seal interface.

The selection of a metal seal in Design Concept 3-1 was dictated by an operational environment of a relatively high temperature (+860°F). This temperature level is prohibitive for most plastics that might otherwise have been suitable for this application. The relatively low wear rating, 6.5, is an average of the ratings (4.8, 4.7, 8.2, and 8.2) for four Design Concept 3-1 type interfaces that were evaluated during the wear analysis. The abnormal wear ratings (4.8 and 4.7) were attributed to heavy shearing and dislocation of gold-plating on the seal faces. Although the exact mechanism of this extensive surface deterioration is unknown, it is suspected to be related to improper gold adherence, possibly as a result of inadequate pretreatment of surfaces before plating. With improved gold plating or application of an alternate sealing surface treatment, these seal interfaces may well achieve an extended or long operational life status in high-temperature applications.

The Mylar and Teflon seal interfaces are similar to those previously reviewed for Interface Category Number 1. Although the seal interfaces of this category (rotary shaft seals) provide a different function than those of the first category (reciprocating shaft seals), the respective comments from the former category apply here also.

4. Rotating Main Closure Seals — Although the major interfaces of this category represent radically differing, rotating main-closure sealing concepts, the only noteworthy variation seen in their assessment concerns the significantly lower wear rating of the metal-to-metal contact sealing interface (Design Concept Number 4-1) in contrast to those interfaces employing a plastic sealing element (Design Concept Numbers 4-2, 4-3, and 4-4).

The metal-to-metal contact sealing interfaces, Design Concept 4-1, are not likely candidates for future applications. Their abnormal wear rating (4.3) was a result of the severe surface deterioration of interface sealing surfaces, observed during wear analysis. Because the operational temperature environments for the component are apparently not prohibitive to the efficacy of the plastic sealing elements, there is nothing to be gained by further evaluation of the existing interfaces at this time.

Each of the sealing interfaces illustrated in Design Concepts 4-2 and 4-4 employs a KEL-F lipseal mating with a steel closure element (butterfly). Except for the obvious similarity of concept, each interface presents subtle differences which, on occasion, suggest limitations to its future usage. The similarities, differences, and assessed limitations of these interfaces are highlighted as follows:

The seals employed in the interfaces of Design Concept 4-2 are machined from KEL-F plate, whereas the seal employed in the interface of Design Concept 4-4 is hot-formed from a KEL-F sheet blank. Because of sporadic fracture problems with the hot-formed seals under low-temperature operation, the machined seals are preferred.

The interfaces of Design Concept 4-2 were designed for low-pressure operation, upstream of the engine turbopump; the interface of Design Concept 4-4 was designed for significantly higher pressure operation, downstream of the engine turbopump. Because of the higher operating pressures imposed on the latter seal, it is thicker and more highly preloaded (with greater diametral stretch on installation over the mating butterfly) than the low-pressure seals. These characteristics suggest that the high-pressure seal may be susceptible to greater wear.

The S-II/100-1-1 seal interface (0.2 in.), illustrated in Design Concept 4-2, has demonstrated long-life operational capability. Larger seal sizes of this configuration, such as the 8-inch seals of Interfaces S-II/89-1-1 and S-II/90-1-1 (Design Concept 4-2), are not recommended for any future applications because their dimensional instability characteristics result in sporadically folded seal lips. Total loss of sealing capability frequently results from progressive seal folding. Less is known about the endurance limit of the 4-inch seal interface illustrated in Design Concept 4-4; however, the interface presently exhibits no adverse characteristics which would exempt it from extended operational service.

All surface finishes used on the seals (8 rms, machined; 32 rms, hot-formed) and the mating butterflies (8 rms) have exhibited no adverse influences on sealing capability; hence, further refinement is unnecessary.

As a product of the times (design phase), A-286 CRES has been a predominant choice for the butterfly closure elements. But its machinability is inherently poor, and we strongly recommend that it be replaced for future seal interface designs. Both from the standpoint of equivalent strength and superior machinability, beryllium copper may be a suitable alternate material.

Design Concept 4-3 is a good candidate for future applications. Although performance anomalies encountered during endurance testing of its using component were directly identified to this seal interface, nevertheless, its design optimization is considered highly promising; briefly, its function will involve optimization of the spring preload-to-pressure balance across the seal interface.

5. and 6. Reciprocating Poppet, Ball, or Piston Seats — For comparison purposes, the plastic seats of Interface Category Number 5 and the metal-to-metal seats of Interface Category Number 6 will be evaluated together. An initial comparison of the two interface categories reveals that the interfaces which contain a plastic member exhibit an overall higher average wear rating than the comparable metal-to-metal interfaces do. With this observation, the question arises as to why metal-to-metal interfaces are used at all. A brief review of design thinking in the past is necessary to answer this question. Primarily, metal-to-metal seats have found application in components under these conditions:

Seat interfaces were loaded beyond the mechanical limits of the available plastics.

Seat interfaces were exposed to high-temperature environments above the recommended operating levels of the available plastics.

With the relatively recent development of the polyimide resins, of which Vespel is a chief representative, and their current ready availability, a major gap between the metallics and the plastics is being bridged. Vespel provides an excellent compromise between the highly tolerant seating alignment characteristics of the softer plastics (Teflon, KEL-F, etc.) and the dimensional and thermal stability of the metallics. From an overall wear standpoint, this material has demonstrated wear resistance superior to any of the other materials in either category. Vespel, for example, has these major attributes:

Vespel has an operating temperature range from -423°F to +900°F. This range adequately covers the operating extremes of all tabulated interfaces in the two categories or, for that matter, any similar seating applications in the foreseeable future.

Its basic resin tensile strength is 13,000 psi at room temperature. This strength is superior to most of the commonly used plastic seating materials; in fact, seats fabricated from this material can operate at seat loads previously prohibitive to plastics.

Vespel's wear characteristics are negligible, as demonstrated in the seat interfaces of two endurance-tested components.

Besides the basic polyimide resin, numerous filled resins (graphite-filled, glass-filled, Teflon-filled, etc.) are available to broaden its utility.

In the future, certain seating applications may still require the extremely low leakage control which is attainable through full implementation of current metal-to-metal seat technology, especially in the small seat sizes. However, none of the seat interfaces in Interface Category Number 6 exhibit this degree of sophistication, nor is such sophistication even warranted. For similar seating applications in the future, the use of polyimide seats should be fully investigated before any consideration of metal-to-metal configurations. In view of this strong recommendation, a detailed evaluation of the relative merits of various metal-to-metal seat interfaces (Interface Category Number 6) is not considered relevant at this time.

The previous recommendations are equally applicable to most of the plastic interfaces of Interface Category Number 5. The superior seating capability and wear resistance of the polyimides were first identified during an assessment of Design Concept 5-11, after the using component had successfully demonstrated long-life operational capability. Subsequently, replacing the seat illustrated in Design Concept 5-8, a polyimide (Vespel) seat was incorporated into a modified valve. During endurance testing of this modified component, the Vespel seat material was demonstrated to be capable of long operational life. In fact, after 10,000 cycles, the polyimide seat in the component exhibited negligible wear and still maintained excellent leakage control. Similarly, polyimide seats were fabricated to replace the seats illustrated in Design Concepts 5-

1 and 5-2. Because of time and funding considerations, however, these seats were not tested in a modified component. The polyimide seats are also considered to be very appealing replacements for most of the other existing seats illustrated by the interfaces of this category.

7. and 8. Reciprocating Guides — All of the piston and shaft guide-bearing interfaces of Interface Category Number 7 have an FEP Teflon or filled-Teflon bearing element. Teflon and the filled versions of this polymer are the favored guide materials because of their extremely low coefficient of friction with mating-member surfaces and resulting low frictional loads. The following guidelines provide a few precautions that must be exercised in the application of these plastic members:

The bearing length to diameter (*L/D*) ratio is critical only in single bearing applications. For example, Design Concept 7-1 illustrates an undesirable *L/D* ratio (0.398) for a single bearing member. In this case, relatively high side-loading will tend to force the shaft to cock about the short guide bearing. This misalignment allows for metal-to-metal contact between the shaft and the retainer guide bore, resulting in severe surface deterioration of these members on either side of the plastic bearing member and in the lowest wear rating (5.7) in this category. In contrast, the single bearing member of Design Concept 7-8 illustrates the highest *L/D* ratio (1.765) of this category, and, in turn, the highest wear rating (9.8), indicative of minimal wear. A bearing *L/D* ratio of at least 1.0 is recommended for all single bearing applications.

For multiple in-line bearing applications, the center-to-center distance between bearings is of greater significance than the *L/D* ratio of the individual bearing members. However, only extremely short center-to-center distances between bearing elements are likely to present any major problem. There is no concern with the multiple bearing interfaces illustrated in Design Concepts 7-2 through 7-7.

The surface finish on the bearing members is apparently not a major issue because it is typically unspecified. The 125-rms surface finish is established by the general callouts on the detail drawing or by the applicable machining specification.

The type of surface finish on the members mating with the plastic bearings is frequently dictated by other functional requirements imposed on the surfaces, such as sealing interfaces in close proximity. In any case, surface finishes in excess of 32 rms are not recommended because of potentially excessive particle generation.

The filled-Teflon materials are particularly suitable for those applications where anticipated high side-loading may tend to extrude bearings fabricated from the basic polymer because of its inherent cold-flow characteristics.

The numerous interfaces in Interface Category Number 8 illustrate the extensive use of metal-to-metal guides. A comparison of these interfaces with the comparable plastic interfaces of Interface Category Number 7 reveals the consistent higher overall average wear rating of the latter. In general, metal-to-metal guides should be avoided in applications where comparable plastic guides can be substituted. For example, serious galling and ultimate seizure problems, associated with the interfaces illustrated in Design Concepts 8-6 and 8-7, were successfully resolved by incorporating polyimide resin (Vespel) guide members in a modified component at these interfaces. In certain future applications, however, the use of plastic guides may not be practical. For those applications, the following guidelines for metal-to-metal guides have been generated through an evaluation of the existing guides illustrated in this category:

The fluted (three-lobed) guides illustrated in Design Concepts 8-16 and 8-17 exhibited the least wear, apparently because of large, effective *L/D* ratios, coupled with limited, distributed loading over the metal-to-metal contact lobe surfaces.

For the other single guide interfaces (Design Concepts 8-1, 8-2, 8-3, 8-4, 8-5, 8-9, and 8 11), the *L/D* ratio appears to have the greatest influence on their wear characteristics. Of these interfaces, those with the highest wear rating also exhibit the highest *L/D* ratio (see Design Concepts 8-9 and 8-11). Correspondingly, the interfaces with the lower *L/D* ratios exhibit the lowest wear ratings in the category. Further, the five lowest wear-rated interfaces (Design Concepts 8-1 through 8-5) illustrate increasing wear rating as a function of decreasing specimen operational cycles.

The effect of lubricants on the observed wear characteristics of the single guide interfaces is identified through a comparison of Design Concepts 8-9 and 8-11, which have comparable *L/D* ratios. By using a lubricant (Microseal Process No. 100-1) between the mating guide members, Design Concept 8-11 exhibits a higher wear rating than the nonlubricated interface of Design Concept 8-9.

Each of the remaining interfaces of this category (Design Concepts 8-8, 8-10, 8-12, 8-13, 8-14, and 8-15) is in line with at least one other interface, providing a guidance function; hence, the *L/D* ratio of these guides is not a critical assessment parameter. Many of the guide interfaces employ the same material for both mating members—and this is generally taboo. However, the adverse effects of mating like-materials have been offset by the use of dry film lubricants or hard-surface platings on one of the mating surfaces at each interface.

There is apparently no logical justification for the sometimes haphazard assignment of surface finishes to the various interface surfaces; hence, a consistent 16 to 32-rms finish is recommended for future applications.

9. Rotating Shaft Journal Bearings — As we expected, the single plastic journal-bearing interface in this category exhibited a higher wear rating than any of the metallic journal bearings. Although this observation was an important finding in the evaluation of previous interface categories, it is of lesser significance in this category because load-carrying capability is the primary selection criterion for journal bearings. Thus where load levels permit, plastic journals are a natural selection over their metallic counterparts. However, as bearing contact loads begin to exceed the support levels of the plastics (Teflon, in this case), metallic journals become a must. This was a true materials selection criterion for the metallic journals of this category. Nevertheless, with the relatively recent development of the higher strength plastics (e.g., the polyimides), plastic journals are expected to find broader application in the future.

The first four journal bearing interfaces of this category (Design Concepts 9-1 through 9-4) represent the same basic configuration—a chrome-plated A-286 CRES or Inconel 718 shaft, rotating in a silver-plated beryllium copper journal bearing. Although this interface configuration represents a successful development, an apparently haphazard assignment of surface finishes was made to the various interfacing surfaces; hence, a consistent 8 to 16-rms finish is recommended for future applications.

The remaining metallic journal-bearing interface, Design Concept 9-5, is conceptually the same as the four preceding interfaces.

10. Rotating Shaft Ball and Roller Bearings and Ball Screws — The bearing units, per se, are a product of a well-established technology which has been fully exploited by the various suppliers. These bearings are normally employed in high-load applications where resulting high frictional loading of simpler journal bearings is unacceptable or undesirable. If these bearings are properly selected and installed for the specific application and applied load levels, their internal bearing surfaces normally reflect negligible wear characteristics. Therefore, the wear analyses were concentrated on the bearing installations, rather than on the internal working surfaces of the bearings themselves. These analyses examined the bearing installations for signs of race slippage as a result of improper retention. Race slippage, as such, is not a major consideration; however, the potential migration of excessive attendant particle generation may adversely affect the bearings themselves or other contaminant-sensitive interfaces which are in close proximity. With few exceptions, the bearing installations of this category exhibited the now-familiar signs of race slippage which are indicative of the lack of positive race retention. For example, the inner and outer race interfaces of the bearing installation, illustrated in Design Concept 10-1, exhibited extensive slippage-induced wear (galling), possibly initiated by vibration-induced fretting, which was also quite evident. As a result, undesirable quantities of loose particulate matter were recovered in the proximity of these bearings. In contrast, the roller-bearing races of Design Concept 10-10 exhibited no signs of slippage, with the corresponding absence of generated particulate matter.

In conclusion, we recommend that all future bearing installations have provisions for positive race retention. The following methods are representative of many for bearing race retention:

High-axial loading with threaded retainers
Radial keying or pinning
Radial press or shrink fits with mating members
Various combinations of the preceding methods

11. Rotating Pin-Joint Bearings — Pin-joint bearings are a critical feature of those components which utilize a linkage system for the transmission or conversion of actuation power. Too often, these bearings have been relegated to a position of secondary consideration, and insufficient attention has been paid to the details of their design. Consequently, the failure of components to operate is frequently traced to pin-joint seizure. In one example, the endurance testing of a component that utilized the pin-joint bearing configuration, illustrated in Design Concept Number 11-1, was terminated after a succession of events indicated increasing frictional resistance within the drive mechanism. The increasing resistance to actuation was traced to impending seizure of the noted pin joint. In turn, the problem was traced further to the utilization of a highly leaded bronze bushing at this interface. Bronze was determined to be a poor selection because it is notorious for breakdown under cryogenic temperature operation.

Many times, pin-joint failures occur as a result of the fact that these mechanisms simply function as a relief feature (a weak link) for more fundamental design deficiencies. Linkage misalignment is considered to be the most common contributor to excessive pin-joint loading and premature seizure of these mechanisms. Such misalignment problems may not be evident during component assembly because they may contribute only to high pin-joint loading (binding) at intermediate or extreme linkage traverse positions. The spherical bearing (self-aligning) pin-joint configurations, depicted in Design Concepts 11-9 and 11-10, were devised to accommodate certain degrees of linkage misalignment. The designs provide for load distribution across the bearing interfaces over the traverse of associated linkage members, thereby significantly increasing the operational longevity of the pin joints.

Primarily because of the lack of a common assessment basis, an in-depth evaluation of each design concept within this category was not undertaken. Each of these interfaces is individually tailored to function within the preset design confines of its using component. However, a few general guidelines about pin-joint design, including some novel features of the tabulated concepts, are highlighted as follows:

The self-aligning, spherical bearing pin-joint configurations of Design Concepts 11-9 and 11-10 are recommended for all future applications where linkage misalignment is anticipated.

If linkage members are not fabricated from materials with inherent bearing qualities, the use of bushings at the pin joints has been found to be effective in reducing wear. For cryogenic applications, beryllium copper bushings are preferred to those fabricated from bronze alloys.

Lubricants, including silver plating and the various baked-on dry film lubricant compounds, have been extensively used in the pin joints with varying degrees of success. A common drawback of lubricating materials concerns their localized breakdown under high unit loading, which is often a result of linkage misalignment. The surface-penetrating lubricant processes, such as the Microseal process, may warrant further investigation for future applications.

Previous surface-finish specifications for pin and mating-member bearing surfaces have appeared to be haphazard; for future applications, a consistent 8 to 16-rms finish is recommended.

Pin retention has been accomplished in a variety of ways. Collaring (as pictured in Design Concept 11-1) or confining the pin between other component elements (e.g., Design Concepts 11-2 and 11-3) are acceptable methods. Design Concept 11-5 shows a novel approach to pin retention. Here, positive pin retention is accomplished through an offset guide-pin interlock. The use of snap-rings (Design Concept 11-11) is the least desirable method of pin retention.

12., 13., 14., and 15. Thrust Bearings, Miscellaneous Guides/Bearings, Stops, and Ball Locks — At present, these interface categories are included only for reference. In each category, the interfaces are loosely bound together only from a functional standpoint. They tend to represent relatively isolated configurations, particular to specific components. Further, these interfaces are not considered to be representative of the more critical aspects influencing component operational longevity. As a result, their collective evaluation is not considered to be particularly meaningful at this time.

Warnings and Predictions of Component Failure

IN today's society, warning lights and acoustical alarms are commonplace. For instance, we are warned by an alarm when our car door does not fully close or when we forget to remove the ignition key. In such instances, we correct the problem instantly — we need no predictions of specific events to follow, nor is an analysis required about the causes: we *know*. Other warning devices that are not as simple — a smoke alarm, a burglar alarm, or the oil-pressure light in our car— raise the specter of disastrous consequences that may follow (or that are already happening): our house burning down; our valuables stolen; or our car engine ruined. The causes though, may not necessarily be apparent.

But wouldn't it be wonderful if there were a device, attached to the components of a rocket engine, which could tell us, with reasonable forewarning, that the component's failure is imminent—rather than just indicating it has *already* failed? Ideally, such a device would transmit a signal that is received by a central monitor, identified, interpreted, and analyzed—with the resultant display on a screen or annunicator to tell the observer *which* component is in trouble, *what* the nature of the anomaly is, and *what to do about it*—or even better, just to tell him *credibly* that all is well.

There have indeed been efforts to develop such systems. Concerning the terminologies used in this field of technology, a term used frequently is "Signature Analysis." The term is derived from the fact that a monitor (such as an accelerometer), attached to a component having moving parts, registers distinct vibrations or sounds when the component parts are in motion. When fed into an oscillograph, the transmitted signals generate a characteristic "signature" trace. The idea is that as long as this trace (analogous to the medical cardiogram of a healthy person) repeats its unchanged pattern, its parts are assumed to be functioning normally. It is also assumed that a change is indicative of deterioration of some kind and that the nature of it can be identified and analyzed by a "comparator," a computer-like device, which does electromechanically what a doctor does when comparing your current electrocardiogram with last year's.

Figure 6-1 displays a typical oscillogram obtained during the actuation (pickup) of a pilot-operated solenoid valve. The recording shows the transmitted signals for voltage and current, and for three accelerometers of different g-load sensitivity, mounted to the valve in the X, Y, and Z axes.

Figure 6-2 displays the tracing of this oscillogram for voltage, current, and two of the accelerometer axes. Time T_1 indicates the seating of the solenoid armature in the pilot (first spike); T_2 indicates the additional time required for the valve to open (second spike).

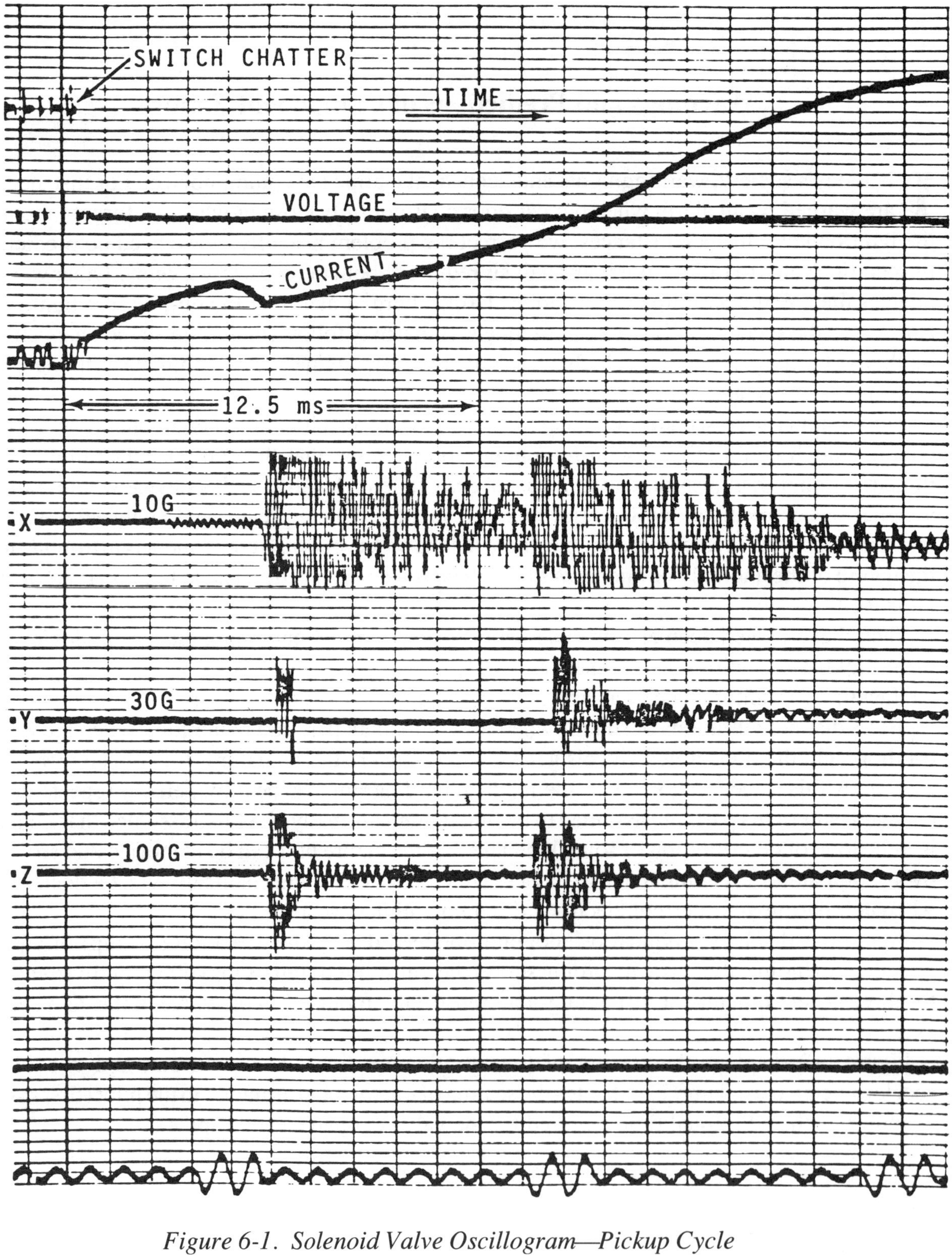

Figure 6-1. Solenoid Valve Oscillogram—Pickup Cycle

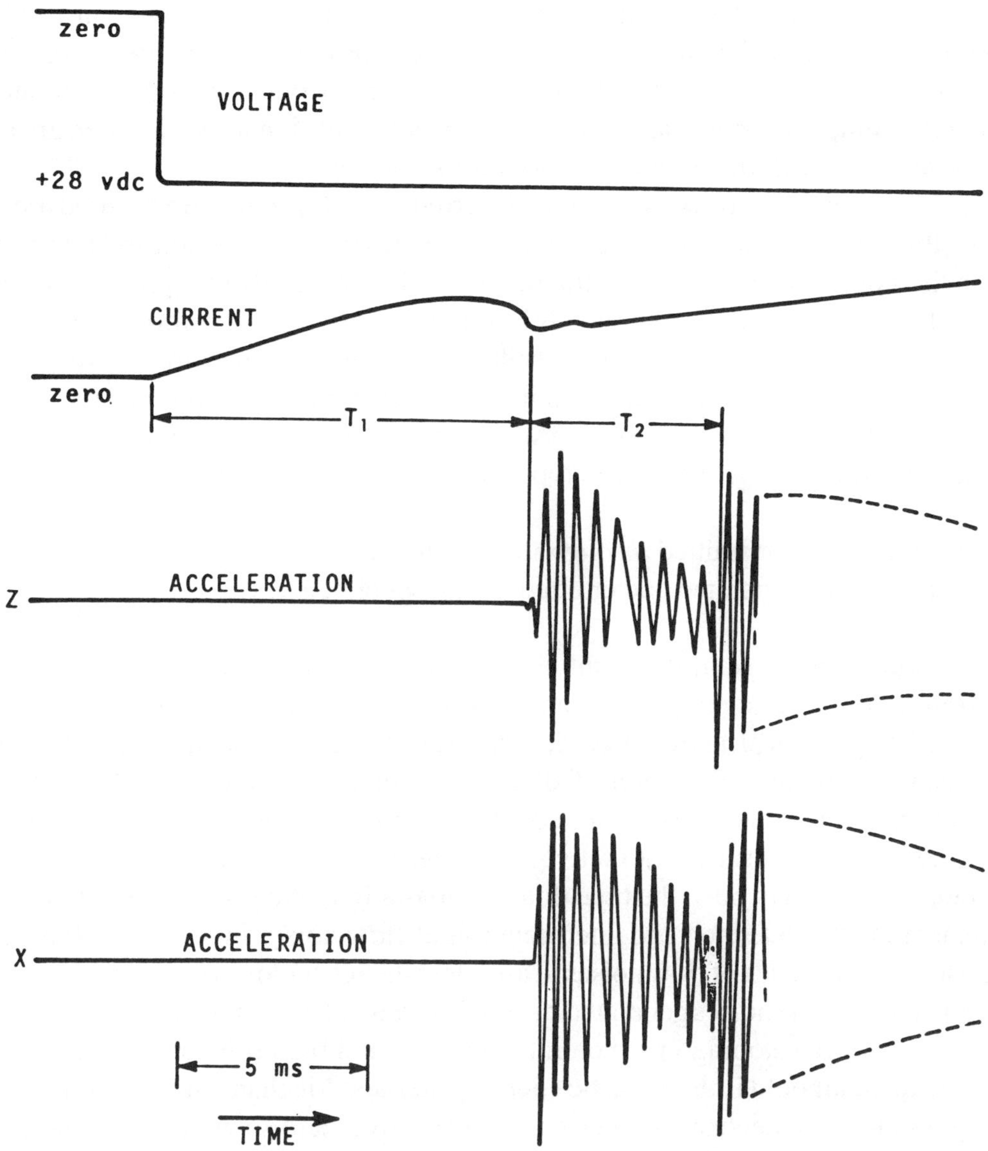

Figure 6-2. Tracing of Solenoid Valve Pickup Voltage, Current, and Accelerometer Oscillograph Records

Signature Analysis Methodologies

There are two principal types of Signature Analysis:

Electrical Signature Analysis—This analysis monitors the current and voltage pickup and decay of a solenoid valve during actuation and dropout. Both parameters would be affected if, for example, the plunger (armature) started to bind because of contamination or deformations affecting the reluctance characteristics of the magnetic circuit.

Mechanical Signature Analysis—This method is the analysis of the output of an accelerometer or microphone (both of which employ essentially the same principles), mounted to the housing of a bearing or the body of a valve. The vibrations and noises (acoustic energy), generated during the motion of the component parts, are converted into electrical signals, which then are processed in the same manner as with the Electrical Signature Analysis.

Mechanical Signature Analysis, sometimes referred to as Structure Born Acoustics, aims at utilizing the vibration characteristics of mechanical devices for the detection and interpretation of anomalies in their operation and/or structural integrity. This method holds promise for both quality control and factory checkout and/or condition-monitoring (fault detection) during operation. In some publications, successes with both procedures have been claimed, mainly for cyclic (rotating) devices, such as bearings. In addition to its use in laboratories, the method has the potential of storing and/or telemetering these recorded signals over long distances for remote data processing and condition-monitoring. Two basic types of signals may be utilized:

1) Acoustic energy transmitted *through* the specimen
2) Acoustic energy generated and emitted *by* the specimen

In Case 1, acoustic energy in the audible or ultrasonic frequency range is applied to, and transmitted through, the (passive) specimen and picked up by suitable transducers (accelerometers, microphones), then processed, and displayed, usually as a frequency spectrum (Fourier spectrum). Typical applications involve detection of discontinuities in structural members (cracks) and imperfect seating of valve elements (wear, contamination, and other mechanical defects).

In Case 2, the vibrations set up by the (active) specimen itself are picked up, recorded, processed, and displayed. Display may be in the time domain or as a frequency spectrum, or both. Typical applications include the detection and interpretation of flow noises in leaking valves and monitoring the absence or presence of flow noises to ascertain the open/closed position of a valve. Other fields of application are the noises generated by bearings and other rotating devices (normal versus abnormal); by cracks developing in structural members; and by loose particles inside of sealed boxes. A more difficult field is the acoustic energy generated Mechanical Signatures Analysis of acyclic (single transient) mechanical devices, such as valves and solenoids, and their abnormal changes.

To the Mechanical Component designer, the latter methodology could become a matter of interest when it is combined, where applicable, with electrical (voltage and current transients) signal analyses.

For rotating devices, American Airlines and General Electric (GE) have done some exploratory work.[4,5] American Airlines applies a two-prong approach: Extension of time between overhauls (TBOs) of jet engines (1) as a result of experience and (2) through the application of a condition-monitored maintenance (CMM) program. CCM is described as promising but only experimental. When applied to cyclic (rotating) machinery, CCM generally holds more promise than when applied to acyclic devices, such as valves.

In the field of acyclic devices, GE appears to have done relatively extensive work. Additionally, in connection with the Long-Life Technology Study, the Space Division of Rockwell International, Inc. has performed considerable research and experimental work.

From an exhaustive literature survey, augmented by numerous consultations with specialists in the field and from reported test results, the following conclusions may be surmised:

Acoustic Energy Transmitted Signature Analysis has merit, especially for factory quality control and the like. For installed components on the ground or in flight, however, its suitability appears in doubt.

Acoustic Energy Generated Signature Analysis of cyclic, rotating machinery may eventually reach some degree of maturity.

Acoustic Energy Generated Signature Analysis of acyclic devices, such as valves, currently appears to be difficult, unreliable, and impractical.

Observations and Conclusions

The preceding statements are based upon the following principal observations and conclusions:

When monitoring, by means of an accelerometer, a ball bearing or other rotating device, many normally identical "signatures" will be obtained in rapid succession (one for each revolution). A recording and/or playback should yield a continuous, regular, repetitive pattern in the time domain. Basically, it should be possible, for instance, to search for, and note the effect of, a flat spot on one of the balls whenever it hits a race, probably in irregular intervals. For a fast-rotating device, such a search could be a tedious process. Actually, however, one only wishes to know *whether* such an irregularity exists and *how* it manifests itself (rather than how often it occurs and in what time pattern). Any irregularity in a composite signature for a given revolution is the result of the presence (or change in intensity) of specific frequencies (of which the signature is composed). Special analyzing equipment can be used to resolve the accelerometer output (signature) into its frequency components and to display a continuously updated (cumulative) average. The irregular components, even if not present in every revolution and even if relatively subtle, can thus be identified when comparing the spectrum obtained with an earlier one of the same bearing or with a "standard spectrum." Mechanical Signature methodology is reported to have been successfully applied for several years in commercial industry (GE), in aviation (American Airlines), and probably in others, as previously mentioned.

A logical thought is that similar benefits might be derived for intermittently operating devices, such as valves. GE, under contract to NASA/KSC, is probably the only experimenter in this field. Although GE has reported some success, attempts to duplicate their results at Rockwell International's Space Division, under contract to NASA/MSFC, have not been successful. The main reasons are given in the following explanations:

In an intermittently operating device, monitored as an assembly, sometimes large numbers of dynamic interfaces are involved. Each of these interfaces produces its own characteristic signature, some of which occur simultaneously, overlapping, or in sequence, during actuation.

We are apparently faced with a number of often drastically different subsignatures which have to be individually analyzed. Analyzers are commercially available which could provide three-dimensional plots of the signatures in an Isoplot format. In this case, the spectrum, as it exists at each instant, can be obtained and plotted against time. Such an analysis might preserve much of the signature data and could allow a more meaningful analysis. It has also been suggested that such analysis could be programmed for execution by using general-purpose digital computer equipment.

It is noted that, depending on the operational systems' characteristics of which the component is a part, or on the characteristics of the test program, only one single actuation may be recorded in a relatively long time—perhaps minutes or sometimes even much longer (as compared to a bearing where, for instance, a thousand signatures are obtained per minute). In the case of the acyclic component, the likelihood of cycle-to-cycle variations, obscuring the subtle data sought for, thus appears much greater. A frequency analysis and display, utilizing a number of "slices" during the operational cycle of devices, such as valves, and then feeding them into an averaging device, can be expected to obscure all subtle symptoms—and it is probably at this point where the principal difficulty for acyclic (nonrotating) devices lies.

The solenoid valve that generated the oscillogram (shown in Figure 6-1) failed without forewarning from the monitored and recorded Signature Analysis data. A total of 9,600 cycles and 544 minutes of vibration in all three axes had been accumulated when this valve failed to actuate while being vibrated in the Z-axis (critical axis). The preceding actuations had not exhibited any abnormal symptoms that were reflected in recordings or observations. The test program had consisted of a combination of cycling at -300°F, room temperature, and -55°F, and of random vibration exposure at -300°F. Upon disassembly, the principal cause for failure was determined to have been excessive wear and particulate generation at the interface between the steel armature and steel solenoid coil housing, resulting in seizure.

Figure 6-2 shows some of the difficulties that may be encountered. If the g-load sensitivity range for the X-axis is adjusted to obtain a recording of "first armature motion," then valve poppet motion is obscured by saturation and excessive flow noise. It is conceivable that fine-tuning of all three accelerometer circuits may produce an acceptable readout; however, this approach raises a question of practicality for operational use.

The operating characteristics of a solenoid valve are largely unaffected by irregularities in the controlled fluid, such as pneumatic pressure, but they can be affected by erratic conditions in the controlling electric system. Such conditions will alter or mask solenoid valve motion irregularities from other causes, mainly because of the difficulties encountered in trying to separate the effects.

The operating characteristics of a valve controlled by a fluid are readily affected by irregularities of that fluid, such as pressure fluctuations. Here, too, the irregularities may mask the cause of valve abnormalities.

High-frequency data tend to show a myriad of resonances. In general, these resonances are typical of the mechanical ringing of the component parts and are not subject to change unless a structural failure occurs. Wear patterns, thus, would probably be in the background, well below the resonance peaks.

Particularly for acyclic devices, complexity increases rapidly with the number of moving elements. In addition, normal scatter (cycle-to-cycle variations) poses serious problems. We must consider the fact that the instrumentation has to be tailor-made for each monitored component or system. For reliable recording of the tell-tale signals, sensor placement, attachment, and orientation are crucial. Experience has shown that correct interpretation of the signatures will require great skill.

If used for very accurate timing purposes, Signature Analysis instrumentation can be valuable for acyclic devices in special situations. It is believed, however, that in most instances, other existing means are adequate. These means include the recordings of the current and voltage transients of solenoid valves or of their position indicator signals and of their pressure-and-flow transients.

Those actively involved in Mechanical Signature Analysis with dynamic (active) devices are clearly divided into two groups: the developers (promoters) and the potential users (customers). The developers usually express considerable optimism about the capabilities of Mechanical Signature Analysis, although they often differ on technical details. By contrast, the users (customers) are generally pessimistic, or cautious at best. Comments range from serious doubts to disappointment. The eventual feasibility of Signature Analysis for cyclic devices under favorable conditions is foreseen by most, but technology for acyclic devices presently is considered to be an optimistic prospect for the distant future.

The seemingly limited experimental work done by relatively few agencies has relied mostly on implanted, known-in-advance deficiencies assumed to represent or simulate wear and other degradation effects. The tests performed at Rockwell International's Space Division seem to be the only ones in which the prediction of impending failure from causes unknown-in-advance was attempted. Figure 6-1 and Figure 6-2 show typical recordings obtained during these tests.

An overall conclusion is that the technology of Mechanical Signature Analysis of dynamic devices for the early detection of incipient component degradations, such as wear, is in its infancy (especially for acyclic devices). The complexity of this method has been underestimated, resulting in overly optimistic predictions and ultimate disappointments. Although all of this may sound pessimistic, the concept of Mechanical Signature Analysis for dynamic devices still appears attractive. The vast gap observed or reported between theory and practice is probably attributable more to the often sweeping, ostentatious, and premature claims made about Mechanical Signature Analysis than to the lack of soundness of the basic idea. Future experiments and subsequent developments should concentrate on real-life applications, instead of depending upon mere feasibility demonstrations that utilize artificial and questionable simulations.

References

[1]"Long-Life Technology Research Based on Saturn V Component Experience," final report, SD 73-SA-0087, Space Division, Rockwell International, Downey, CA, July 1973.

[2]"Analysis of Surveyor III Materials and Photos Returned by Apollo 12," 1972, Stock No. 3300-0439, Superintendent of Documents, U.S. Printing Office, Washington, DC, 20402.

[3]Apollo Expeditions to the Moon," edited by Edgar M. Cartright, Scientific and Technical Information Office, National Aeronautics and Space Administration, 1975, NASA SP 350, Superintendent of Documents, Stock No. 033-000-00630-6.

[4]"Mechanical Systems Readiness Assessment and Performance Monitoring Study," K. A. Smith et al., General Electric Company, Schenectady, NY, final report, May 1972.

[5]"Condition Monitored Maintenance" (CMM), information booklet, American Airlines, Inc., undated.

Interface Design Concept Tabulation

THIS tabulation was generated during the wear analysis of 13 mechanical/electromechanical liquid-propellant rocket engine and vehicle components. The components included five cryogenic propellant shutoff valves, three pneumatic system solenoid valves, three pneumatic actuation system regulators or propellant tank pressure regulators, one propellant tank vent and relief valve, and one propellant directional vent valve.

The numbers at the top of each column are a combination of the interface category number and, within each category, the design concept number.

Note that the entries "S-IC," "S-II," and "S-IVB" refer to the first, second, and third stages of the successful Apollo/Saturn moon rocket.

Interface Category Number 1 (Sheet 1 of 2)

RECIPROCATING SHAFT SEALS - Radial Sealing

	DESIGN CONCEPT NUMBER	1-1	1-2	1-3	1-4
COMPONENT DATA BOOK	CONTROL NUMBER(S)	S - II / 89 - 1	S - II / 92 - 3	J - 2 / 208 - 1	J - 2 / 208 - 1
	INTERFACE DASH NUMBER(S)	- 13	- 11	- 17 & - 17A	- 13 & - 13A
	WEAR ANALYSIS RATING (AVG.)	**4.5**	**6.8**	**7.8**	**8.2**
	PHOTOGRAPHS, PAGES	B - 4, - 5, - 13, & - 14	86, 100, & 101	B - 7, & - 21	B - 5, & -17
INTERFACE DESCRIPTION	INTERFACE PICTORIAL DESCRIPTION	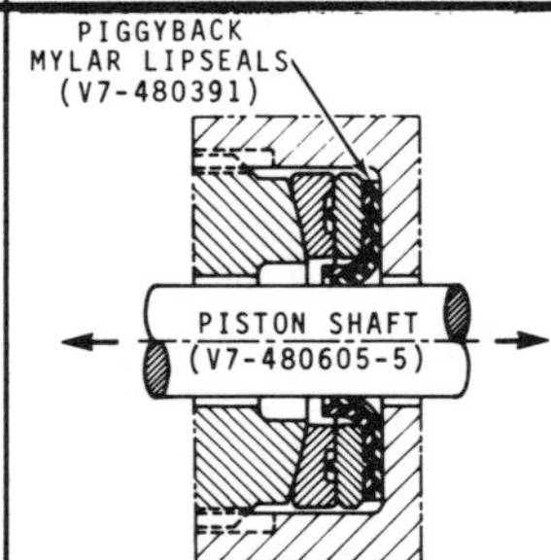	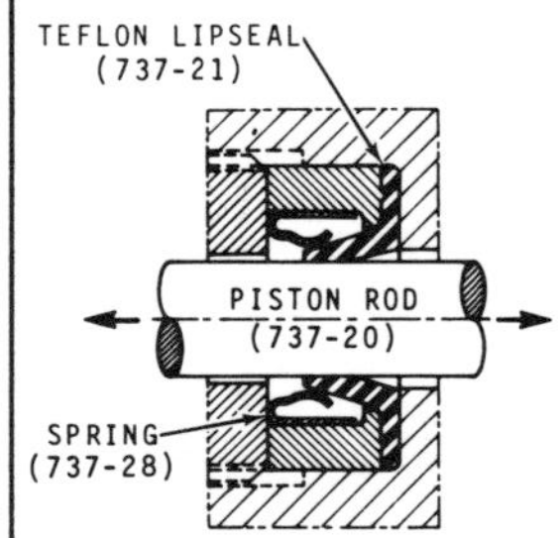	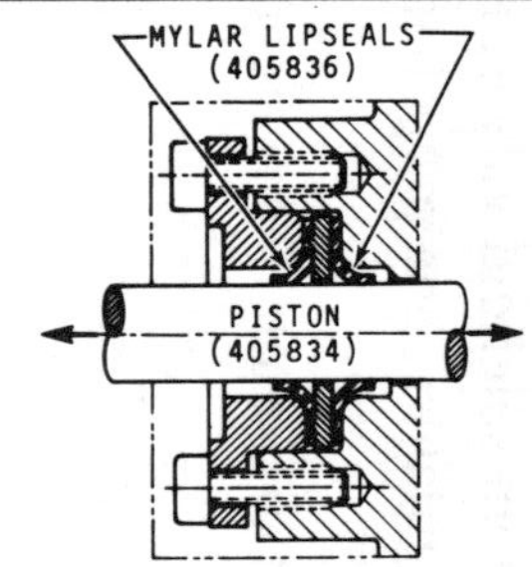	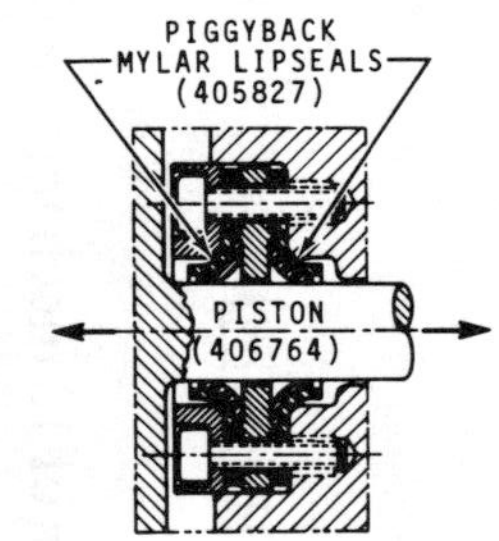
	SEAL TYPE	LIPSEAL	LIPSEAL	LIPSEAL	LIPSEAL
	BASIC SEALING DIA., IN.	.9370	.3740	.250	1.000
	DESIGN OPERATING MEDIA	GHe	GHe	GHe	GHe
	DESIGN OPERATING PRESSURE, PSIG	750	0 to 725	400	400
	DESIGN OPERATING TEMP. RANGE, °F	- 423 to + 160	- 423 to + 160	- 420 to + 140	- 420 to + 140
	MAXIMUM ALLOWABLE LEAKAGE	100 SCIM GHe @ - 320 °F	NO REQUIREMENT	100 SCIM GHe @ - 320 °F	100 SCIM GHe @ - 320 °F
	DESIGN OPERATING LIFE, CYCLES	750	600	1308	1308
	SPECIMEN OPERATION, CYCLES	591	1257	292	292
INTERFACE MEMBERS DESCRIPTION — SEAL	PART NUMBER	V7 - 480391 (2)	737 - 21	405836 (2)	405827 (4)
	PART NAME	LIPSEAL, ROD - FILL VALVE	SEAL - VENT PISTON ROD	SEAL, LIP, SEQUENCE VALVE	LIPSEAL, PISTON ROD - MPV
	MATERIAL	MB0130 - 060 MYLAR	AMS3651 TEFLON	RB0130 - 039 MYLAR	RB0130 - 039 MYLAR
	SURFACE FINISH	AS REC'D. PLASTIC FILM	SMOOTH MACHINE FINISH	AS REC'D. PLASTIC FILM	AS REC'D PLASTIC FILM
	FORMING METHOD	COLD - FORMED	MACHINED	COLD - FORMED	COLD - FORMED
	ADDITIONAL LOADING MEMBER (S)	PIGGYBACK SEALS	737 - 28 GARTER SPRING (BeCu)	NONE	PIGGYBACK SEALS
	BACK-UP SUPPORT MEMBER(S)	NONE	NONE	NONE	NONE
INTERFACE MEMBERS DESCRIPTION — MATING MEMBER	PART NUMBER	V7 - 480605 - 5	737 - 20	405834	406764
	PART NAME	SHAFT, PISTON - PREVALVE	ROD - VENT ACTUATING PISTON	PISTON, SEQUENCE VALVE	PISTON, MAIN PROP. VALVE
	MATERIAL	MB0170 - 041 INCONEL 718	MIL - S - 7720 18 -8 / 303S CRES	AMS5735 A - 286 CRES	A - 286 CRES
	SURFACE FINISH	8 r m s	16 r m s	8 -16 r m s	6 r m s
	SURFACE TREATMENT	THIN DENSE CHROME PLATE	HARD CHROME PLATE	NONE	NONE
	OTHER SURFACE TREATMENTS	PASSIVATE	PASSIVATE	NONE	NONE
USING COMPONENT DESCRIPTION	PROCUREMENT DOCUMENT(S) OR P/N(S)	V7 - 480450 - 121	ME284 - 0030 - 0039	409920X7	409920X7
	COMPONENT NAME	8 - INCH PROPELLANT FILL & DRAIN VALVE	7 - INCH LH_2 TANK VENT & RELIEF VALVE	4 - INCH NORMALLY CLOSED MAIN FUEL VALVE	4 - INCH NORMALLY CLOSED MAIN FUEL VALVE
	MANUFACTURER	ROCKWELL / LAD	AMETEK / CALMEC	ROCKWELL / RKD	ROCKWELL / RKD
	MANUFACTURER'S PART NUMBER	V7 - 480450 - 121	737 - 555	409920X7	409920X7
	USING VEHICLE(S) AND STAGE(S)	SATURN S - II	SATURN S - II	SATURN S - II & SATURN S - IVB	SATURN S - II & SATURN S - IVB
	USING SYSTEM(S)	LOX & LH_2 TANK FILL & DRAIN	LH_2 TANK VENT	J - 2 ENGINE LH_2 FEED	J - 2 ENGINE LH_2 FEED

Interface Category Number 1
(Sheet 2 of 2)

RECIPROCATING SHAFT SEALS
- Radial Sealing

	DESIGN CONCEPT NUMBER	1–5	1–6	1–7	1–8
COMPONENT DATA BOOK	CONTROL NUMBER(S)	S – II / 92 – 3	S – II / 104 – 2; S – IC / 1 – 2	S – II / 100 – 1	
	INTERFACE DASH NUMBER(S)	– 8 & – 8A	– 3 & – 6; – 2 & – 6	– 7	
	WEAR ANALYSIS RATING (AVG.)	**8.5**	**8.9**	**9.6**	
	PHOTOGRAPHS, PAGES	87, & 88	B – 2, – 3, – 6, – 8, & – 9; B – 2, – 3, – 5, & – 6	82, 88, & 91	
INTERFACE DESCRIPTION	INTERFACE PICTORIAL DESCRIPTION	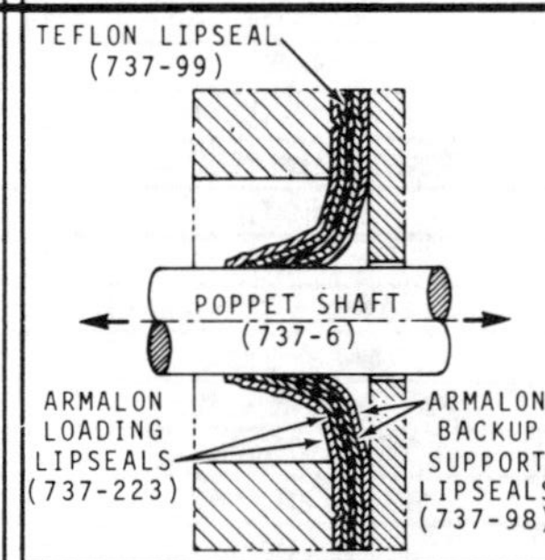	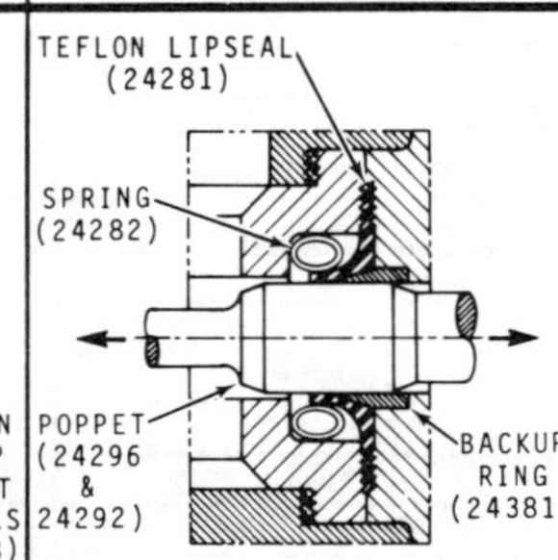	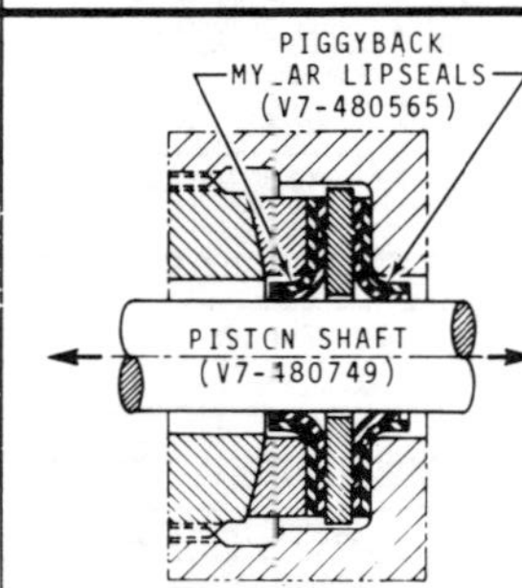	
	SEAL TYPE	LIPSEAL	LIPSEAL	LIPSEAL	LIPSEAL
	BASIC SEALING DIA., IN.	1.7490	.3740	.562	
	DESIGN OPERATING MEDIA	GH_2	GHe	LOX or LH_2	
	DESIGN OPERATING PRESSURE, PSIG	42 PSIA	750	50	
	DESIGN OPERATING TEMP. RANGE, °F	– 423 to + 160	– 180 to + 160	– 423 to + 160	
	MAXIMUM ALLOWABLE LEAKAGE	NO REQUIREMENT	20 SCIM GHe	25 SCIM GHe @ – 320 °F	
	DESIGN OPERATING LIFE, CYCLES	600	2000	750	
	SPECIMEN OPERATION, CYCLES	1257	1096	935	
INTERFACE MEMBERS DESCRIPTION — SEAL	PART NUMBER	737 – 98	24281	V7 – 430565 (4)	
	PART NAME	SEAL – MAIN SHAFT	SEAL	SEAL, PISTON SHAFT – RECIR. VALVE	
	MATERIAL	AMS3651 TEFLON	TECFLUORFIL B (FILLED TEFLON)	MB0130–060 MYLAR	
	SURFACE FINISH	SMOOTH MACHINE FINISH	32 r m s	AS REC'D. PLASTIC FILM	
	FORMING METHOD	COLD – FORMED	MACHINED	COLD – FORMED	
	ADDITIONAL LOADING MEMBER (S)	737 – 223 SEAL (ARMALON # 406 – 116)	24282 SPRING	PIGGYBACK SEALS	
	BACK–UP SUPPORT MEMBER(S)	737 – 99 SEAL (ARMALON # 406 – 116)	24381 RING	NONE	
MATING MEMBER	PART NUMBER	737 – 6	24296 & 24292	V7 – 480749	
	PART NAME	SHAFT – MAIN POPPET	UPPER & LOWER POPPETS	SHAFT, PISTON – RECIR. VALVE	
	MATERIAL	MIL – S – 7720 18 – 8 / 303S CRES	QQ – S – 763 303 CRES	QQ – C – 530 BERYLLIUM COPPER	
	SURFACE FINISH	8 r m s	8 r m s	8 r m s	
	SURFACE TREATMENT	HARD CHROME PLATE	ELECTROPOLISH	THIN DENSE CHROME PLATE	
	OTHER SURFACE TREATMENTS	PASSIVATE	FREE OF NICKS, SCRATCHES, ETC.	PASSIVATE	
USING COMPONENT DESCRIPTION	PROCUREMENT DOCUMENT(S) OR P/N(S)	ME284 – 0030 – 0039	ME284 – 0356 – 0001; 60B52101 – 15F	V7 – 480328 – 91	
	COMPONENT NAME	7 – INCH LH_2 TANK VENT & RELIEF VALVE	3 / 8 – INCH NORMALLY CLOSED SOLENOID VALVE	2 – INCH NORMALLY OPEN RECIRCULATION VALVE	
	MANUFACTURER	AMETEK / CALMEC	J. C. CARTER	ROCKWELL / LAD	
	MANUFACTURER'S PART NUMBER	737 – 555	6969 – 2; 6969 – 15	V7 – 480328 – 91	
	USING VEHICLE(S) AND STAGE(S)	SATURN S – II	SATURN S – II; SATURN S – IC	SATURN S – II	
	USING SYSTEM(S)	LH_2 TANK VENT	VALVE ACTUATION, He INJECTION, & ACCUMULATOR FILL	LOX & LH_2 RECIRCULATION	

Interface Category Number 2
(Sheet 1 of 3)

RECIPROCATING PISTON OR POPPET SEALS
- Radial Sealing

			2-1	2-2	2-3	2-4
DESIGN CONCEPT NUMBER			2-1	2-2	2-3	2-4
COMPONENT DATA BOOK		CONTROL NUMBER(S)	S - IVB / 210 - 3	S - IC / 206 - 1	S - II / 92 - 3	S - IC / 207 - S
		INTERFACE DASH NUMBER(S)	- 3 & - 3A	- 9	- 2	- 2
		WEAR ANALYSIS RATING (AVG.)	**5.6**	**7.0**	**7.6**	**7.9**
		PHOTOGRAPHS, PAGES	B - 3, - 4, - 5, & - 10	B - 4, - 10, & - 11	89, 90, 91, 92, & 93	B - 5, - 6, & - 13
INTERFACE DESCRIPTION		INTERFACE PICTORIAL DESCRIPTION	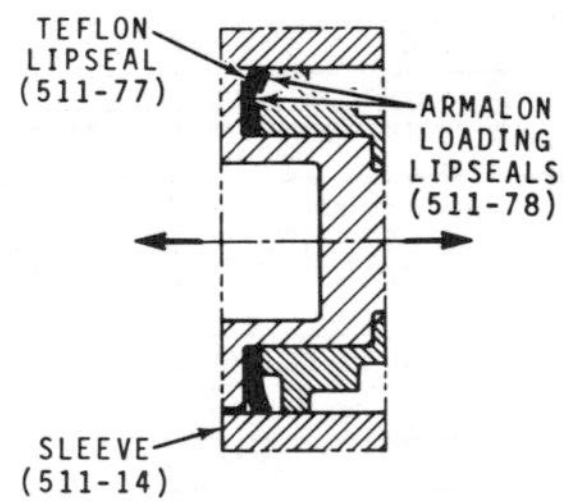	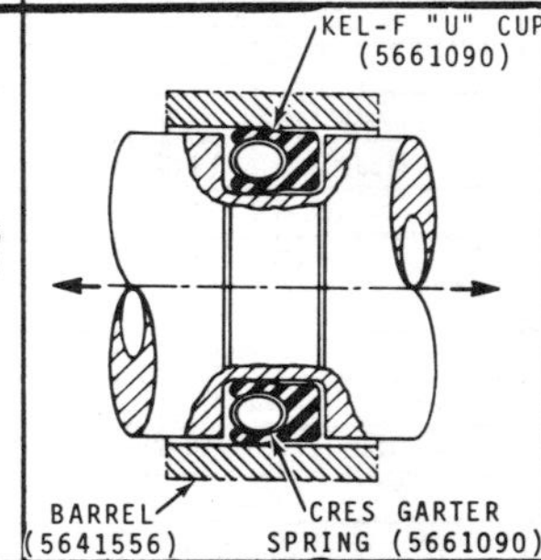	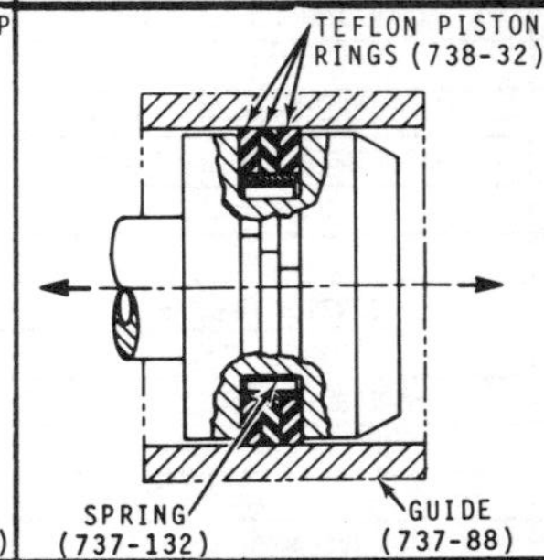	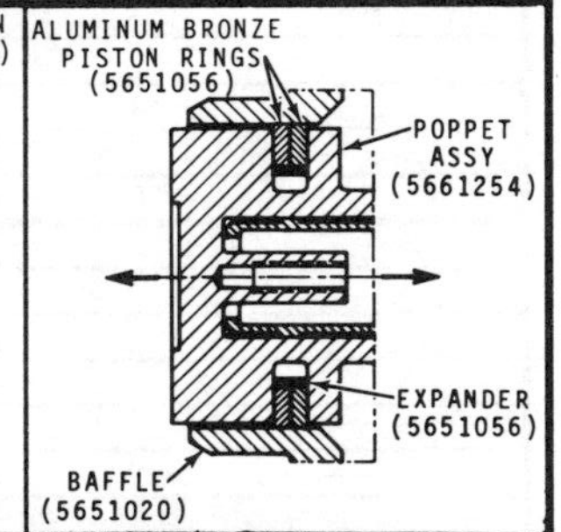
		SEAL TYPE	LIPSEAL	"U" - CUP	PISTON - RING	PISTON - RING
		BASIC SEALING DIA., IN.	.813	2.500	2.188	1.5630
		DESIGN OPERATING MEDIA	GHe	GHe	GHe or GH_2	GHe, GH_2 or AIR
		DESIGN OPERATING PRESSURE, PSIG	475	500 to 750	0 to 32.5	3200
		DESIGN OPERATING TEMP. RANGE, °F	- 423 to + 160	- 423 to + 165	- 423 to + 160	- 400 to + 165
		MAXIMUM ALLOWABLE LEAKAGE	50 SCCM	GROSS LEAKAGE CONTROL ONLY	NO REQUIREMENT	GROSS LEAKAGE CONTROL ONLY
		DESIGN OPERATING LIFE, CYCLES	200	2000	600	500
		SPECIMEN OPERATION, CYCLES	30	3318	1257	500
INTERFACE MEMBERS DESCRIPTION	SEAL	PART NUMBER	511 - 77	5661090	738 - 32 (3)	5651056 (2)
		PART NAME	SEAL - LOCK PISTON	SEAL - DYNAMIC	RING - SECONDARY P.V. PISTON	SEAL, BAFFLE - BALL SCREW
		MATERIAL	AMS3651 TEFLON	KEL - F	AMS3651 TEFLON	ALUMINUM BRONZE
		SURFACE FINISH	SMOOTH MACHINE FINISH	NOT SPECIFIED	16 rms	NOT SPECIFIED
		FORMING METHOD	COLD - FORMED	MACHINED	MACHINED	MACHINED
		ADDITIONAL LOADING MEMBER (S)	511 - 78 SEAL (2) (ARMALON 410 - 128)	INTEGRAL CRES GARTER SPRING	737 - 132 SPRING	EXPANDER (HAYNES 25 STELLITE)
		BACK-UP SUPPORT MEMBER(S)	NONE	NONE	NONE	NONE
	MATING MEMBER	PART NUMBER	511 - 14	5641556	737 - 88	5651020
		PART NAME	SLEEVE - BALL RETAINING	BARREL - ACTUATOR	GUIDE - SECONDARY PILOT VALVE	BAFFLE, BALL SCREW
		MATERIAL	AMS5643 17 - 4PH CRES	QQ - A - 200 / 8 6061 - T651 AL. ALY.	QQ - A - 325b 6061 - T6 AL. ALY.	AMS5737 A - 286 CRES
		SURFACE FINISH	32 r m s	32 r m s	10 r m s	8 r m s
		SURFACE TREATMENT	PASSIVATE	HARD ANODIZE	MIL - A - 8625 ANODIZE	MICROPLATE NO. 1
		OTHER SURFACE TREATMENTS	NONE	NONE	NONE	PASSIVATE
USING COMPONENT DESCRIPTION		PROCUREMENT DOCUMENT(S) OR P/N(S)	1A49988	20M32014	ME284 - 0030 - 0039	20M32015
		COMPONENT NAME	6 - INCH DIRECTIONAL CONTROL FUEL VENT VALVE	2 - INCH NORMALLY CLOSED SUBMERGED LH_2 SHUTOFF VALVE	7 - INCH LH_2 TANK VENT & RELIEF VALVE	1 - 1 / 2 INCH HELIUM PRESSURE REGULATOR
		MANUFACTURER	AMETEK / CALMEC	PARKER - HANNIFIN	AMETEK / CALMEC	PARKER - HANNIFIN
		MANUFACTURER'S PART NUMBER	511 - 509	5640027	737 - 555	5640061
		USING VEHICLE(S) AND STAGE(S)	SATURN S - IVB	SATURN S - IC (QUALIFIED BACK - UP)	SATURN S - II	SATURN S - IC (QUALIFIED BACK - UP)
		USING SYSTEM(S)	LH_2 TANK VENT	DNA	LH_2 TANK VENT	DNA

Interface Category Number 2
(Sheet 2 of 3)

RECIPROCATING PISTON OR POPPET SEALS
- Radial Sealing

		DESIGN CONCEPT NUMBER	2-5	2-6	2-7	2-8
COMPONENT DATA BOOK		CONTROL NUMBER(S)	J-2/208-1	S-IVB/210-3	S-II/92-3	S-II/92-3
		INTERFACE DASH NUMBER(S)	-15 & -15A	-7 & -7A	-10	-14 & -14A
		WEAR ANALYSIS RATING (AVG.)	**8.0**	**8.2**	**8.5**	**8.6**
		PHOTOGRAPHS, PAGES	B-6, -18, & -19	B-3, -4, -5, -11, & -12	86	87, & 104
INTERFACE DESCRIPTION		INTERFACE PICTORIAL DESCRIPTION	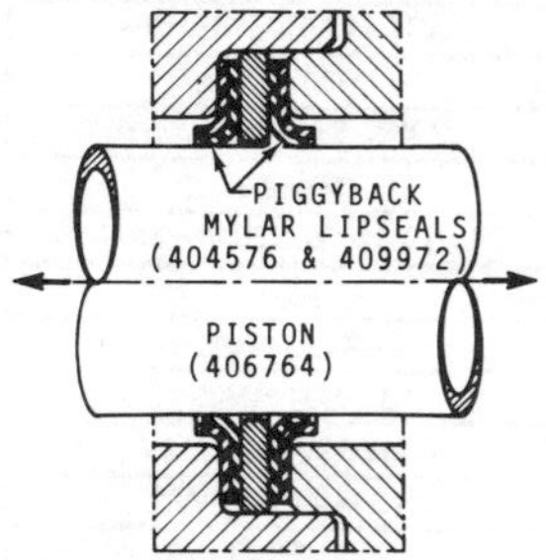	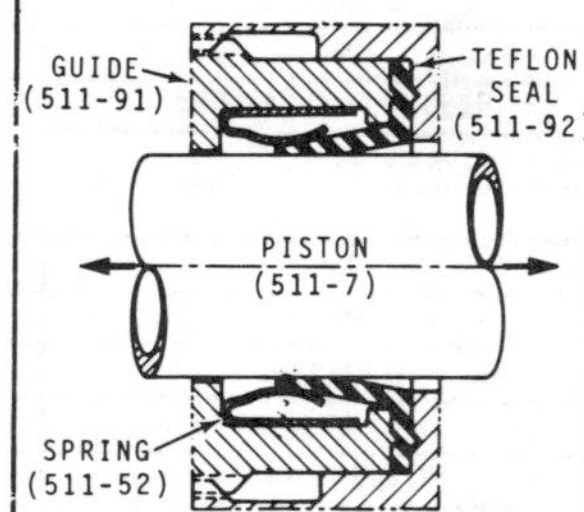	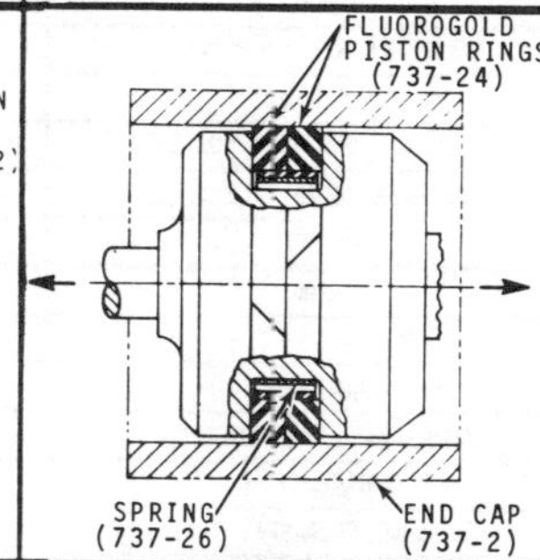	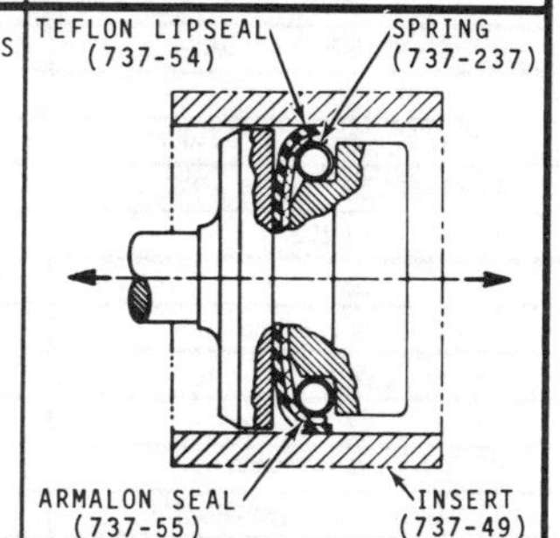
		SEAL TYPE	LIPSEAL	LIPSEAL	PISTON-RING	LIPSEAL
		BASIC SEALING DIA., IN.	3.5300	1.1250	1.750	.892
		DESIGN OPERATING MEDIA	GHe	GHe, GH_2, or LH_2	GHe	GHe
		DESIGN OPERATING PRESSURE, PSIG	400	475	0 to 750	0 to 750
		DESIGN OPERATING TEMP. RANGE, °F	-420 to +140	-423 to +160	-423 to +160	-423 to +160
		MAXIMUM ALLOWABLE LEAKAGE	100 SCIM GHe @ -320°F	50 SCCM	2.5 SCFM, -320°F to -423°F	6.0 SCFM, -320°F to -423°F
		DESIGN OPERATING LIFE, CYCLES	1308	200	600	600
		SPECIMEN OPERATION, CYCLES	292	30	1257	1257
INTERFACE MEMBERS DESCRIPTION	SEAL	PART NUMBER(S)	404576 & 409972	511-92	737-24 (2)	737-54
		PART NAME	SEAL, LIP, ACTUATOR-MPV	SEAL, CONTROL PISTON	RING-VENT ACT. PISTON HD.	SEAL-OVERRIDE PISTON HEAD
		MATERIAL	RB0130-039 MYLAR	AMS3651 TEFLON	FLUOROGOLD	AMS3651 TEFLON
		SURFACE FINISH	AS REC'D PLASTIC FILM	SMOOTH MACHINE FINISH	SMOOTH MACHINE FINISH	SMOOTH MACHINE FINISH
		FORMING METHOD	COLD-FORMED	MACHINED	MACHINED	COLD-FORMED
		ADDITIONAL LOADING MEMBER (S)	PIGGYBACK SEALS	511-52 SPRING-MAIN SHAFT SEAL (BeCu)	737-26 SPRING	737-55 SEAL (ARMALON); 737-55 SPRING
		BACK-UP SUPPORT MEMBER(S)	NONE	NONE	NONE	NONE
	MATING MEMBER	PART NUMBER	406764	511-7	737-2	737-49
		PART NAME	PISTON, MAIN PROPELLANT VALVE	PISTON, CONTROL	END CAP-7" INLINE V/R VALVE	INSERT-OVERRIDE PISTON
		MATERIAL	A-286 CRES	AMS5643 17-4PH CRES	MB0170-030 A356-T6 AL. ALY.	MIL-S-7720 18-8/303 CRES
		SURFACE FINISH	6 r m s	6 r m s	16 r m s	16 r m s
		SURFACE TREATMENT	NONE	HARD CHROME PLATE	MIL-A-8625 ANODIZE	PASSIVATE
		OTHER SURFACE TREATMENTS	NONE	PASSIVATE	NONE	NONE
USING COMPONENT DESCRIPTION		PROCUREMENT DOCUMENT(S) OR P/N(S)	409920X7	1A49988	ME284-0030-0039	ME284-0030-0039
		COMPONENT NAME	4-INCH NORMALLY CLOSED MAIN FUEL VALVE	6-INCH DIRECTIONAL CONTROL FUEL VENT VALVE	7-INCH LH_2 TANK VENT & RELIEF VALVE	7-INCH LH_2 TANK VENT & RELIEF VALVE
		MANUFACTURER	ROCKWELL/RKD	AMETEK/CALMEC	AMETEK/CALMEC	AMETEK/CALMEC
		MANUFACTURER'S PART NUMBER	409920X7	511-509	737-555	737-555
		USING VEHICLE(S) AND STAGE(S)	SATURN S-II & SATURN S-IVB	SATURN S-IVB	SATURN S-II	SATURN S-II
		USING SYSTEM(S)	J-2 ENGINE LH_2 FEED	LH_2 TANK VENT	LH_2 TANK VENT	LH_2 TANK VENT

Interface Category Number 2
(Sheet 3 of 3)

RECIPROCATING PISTON OR POPPET SEALS – Radial Sealing

		DESIGN CONCEPT NUMBER	2-9	2-10	2-11	2-12
COMPONENT DATA BOOK		CONTROL NUMBER(S)	S – II / 100 – 1; S – II / 89 – 1	S – IVB / 209 – 2	S – II / 92 – 3	S – II / 92 – 3
		INTERFACE DASH NUMBER(S)	– 9; – 16, & – 16A	– 12	– 9	– 4
		WEAR ANALYSIS RATING (AVG.)	**8.7**	**9.1**	**9.3**	**9.5**
		PHOTOGRAPHS, PAGES	82; B – 5, – 6, – 16, – 17, & – 18	B6, & – 16	87, 98, & 99	89
INTERFACE DESCRIPTION		INTERFACE PICTORIAL DESCRIPTION	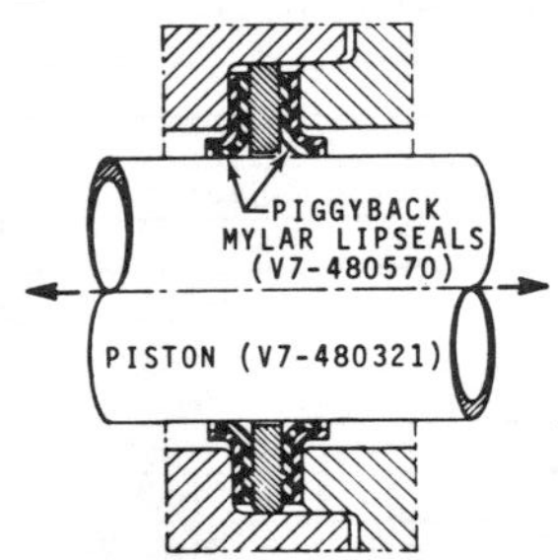	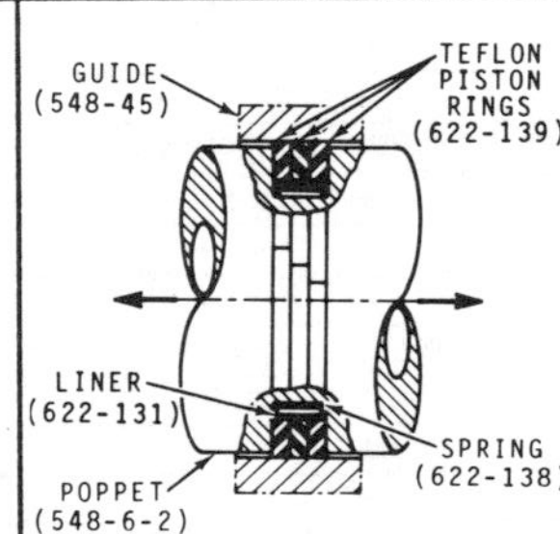	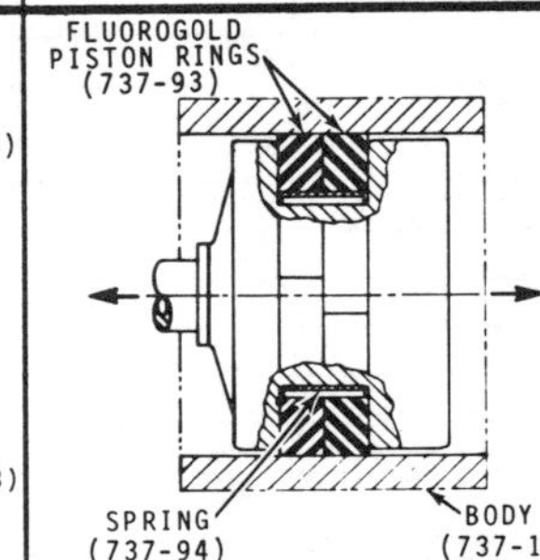	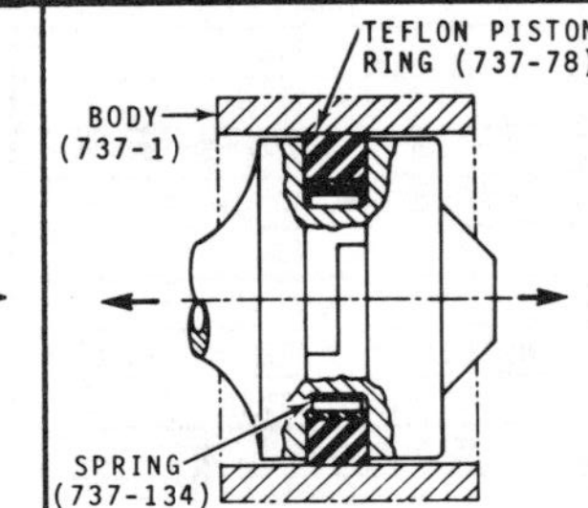
		SEAL TYPE	LIPSEAL	PISTON – RING	PISTON – RING	PISTON – RING
		BASIC SEALING DIA., IN.	1.750; 3.000	.701	9.540	1.292
		DESIGN OPERATING MEDIA	GHe	GHe	GHe or GH2	GHe or GH2
		DESIGN OPERATING PRESSURE, PSIG	0 to 750	25 to 3100	0 to 32.5	0 to 32.5
		DESIGN OPERATING TEMP. RANGE, °F	–423 to + 160	–423 to + 160	–423 to + 160	–423 to + 160
		MAXIMUM ALLOWABLE LEAKAGE	30 SCIM GHe; 500 SCIM GHe	GROSS LEAKAGE CONTROL ONLY	NO REQUIREMENT	NO REQUIREMENT
		DESIGN OPERATING LIFE, CYCLES	750	5000	600	600
		SPECIMEN OPERATION, CYCLES	935; 531	10,121	1257	1257
INTERFACE MEMBERS DESCRIPTION	SEAL	PART NUMBER(S)	V7 – 480570; V7 – 480453	622 – 139 (3)	737 – 93 (2)	737 – 78
		PART NAME	LIPSEALS	SEAL, POPPET SHUTOFF VALVE	RING – MAIN PISTON SEAL	PISTON RING – SECONDARY P. V.
		MATERIAL	MB0130 – 060 MYLAR	AMS3651 TEFLON	FLUOROGOLD	AMS3651 TEFLON
		SURFACE FINISH	AS REC'D PLASTIC FILM	16 r m s	SMOOTH MACHINE FINISH	16 r m s
		FORMING METHOD	COLD – FORMED	MACHINED	MACHINED	MACHINED
		ADDITIONAL LOADING MEMBER (S)	PIGGYBACK SEALS	622 – 138 SPRING – POPPET	737 – 94 SPRING	737 – 134 SPRING
		BACK–UP SUPPORT MEMBER(S)	NONE	NONE	NONE	NONE
	MATING MEMBER	PART NUMBER	V7 – 480321; V7 – 480647	548 – 45	737 – 1	737 – 1
		PART NAME	PISTONS	GUIDE, VALVE POPPET	BODY – 7 " INLINE V / R VALVE	BODY – 7 " INLINE V / R VALVE
		MATERIAL	AMS5737 A286 CRES, 304L CRES	AMS5643 17 – 4PH CRES	MB0170 – 030 A356 – T6 AL. ALY.	MB0170 – 030 A356 – T6 AL. ALY.
		SURFACE FINISH	8 r m s; 4 r m s	8 r m s	63 r m s	16 r m s
		SURFACE TREATMENT	THIN DENSE CHROME PLATE	PASSIVATE	MIL – A – 8625 ANODIZE	MIL – A – 8625 ANODIZE
		OTHER SURFACE TREATMENTS	PASSIVATE	NONE	NONE	NONE
USING COMPONENT DESCRIPTION		PROCUREMENT DOCUMENT(S) OR P/N(S)	V7 – 480328 – 91; V7 – 480450 – 121	1B43660 – 515	ME284 – 0030 – 0039	ME284 – 0030 – 0039
		COMPONENT NAME	2 – INCH RECIRCULATION VALVE; 8 – INCH FILL & DRAIN VALVE	1 / 2 – INCH NORMALLY CLOSED HELIUM SOLENOID VALVE	7 – INCH LH_2 TANK VENT & RELIEF VALVE	7 – INCH LH_2 TANK VENT & RELIEF VALVE
		MANUFACTURER	ROCKWELL / LAD	AMETEK / CALMEC	AMETEK / CALMEC	AMETEK / CALMEC
		MANUFACTURER'S PART NUMBER	V7 – 480328 – 91; V7 – 480450 – 121	548 – 515	737 – 555	737 – 555
		USING VEHICLE(S) AND STAGE(S)	SATURN S – II	SATURN S – IVB	SATURN S – II	SATURN S – II
		USING SYSTEM(S)	LOX & LH_2 RECIRCULATION; LOX & LH_2 FILL & DRAIN	PRESSURE CONTROL	LH_2 TANK VENT	LH_2 TANK VENT

Interface Category Number 3
(Sheet 1 of 1)

ROTATING SHAFT SEALS
– Radial or Face Sealing

		DESIGN CONCEPT NUMBER	3-1	3-2	3-3	3-4
COMPONENT DATA BOOK		CONTROL NUMBER(S)	S – II / 98 – 5	S – IVB / 210 – 3	J – 2 / 208 – 1	S – II / 89 – 1
		INTERFACE DASH NUMBER(S)	– 15 & – 15A	– 18	– 3 & – 4; – 3A & – 4A	– 4, – 4A, – 5, & – 5A
		WEAR ANALYSIS RATING (AVG.)	**6.5**	**7.4**	**8.0**	**8.3**
		PHOTOGRAPHS, PAGES	B – 3, – 4, & – 19	B – 8, & – 19	B – 2, – 3, – 4, – 11, – 12, – 13, & – 14	B – 2, – 3, – 8, & – 9
INTERFACE DESCRIPTION		INTERFACE PICTORIAL DESCRIPTION	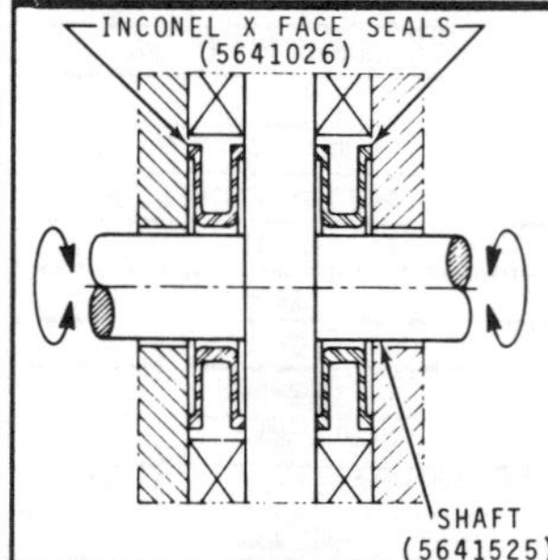	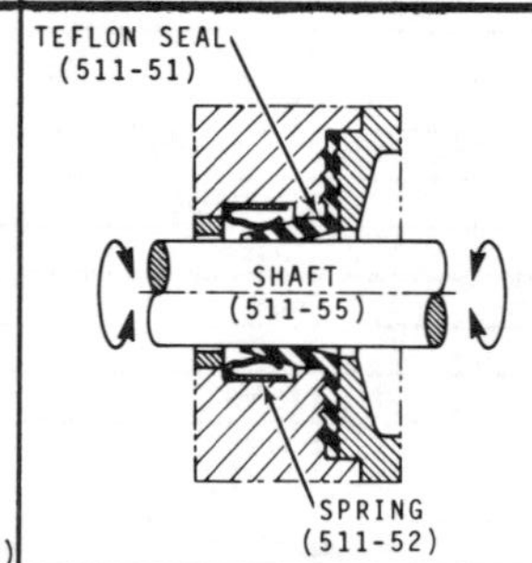	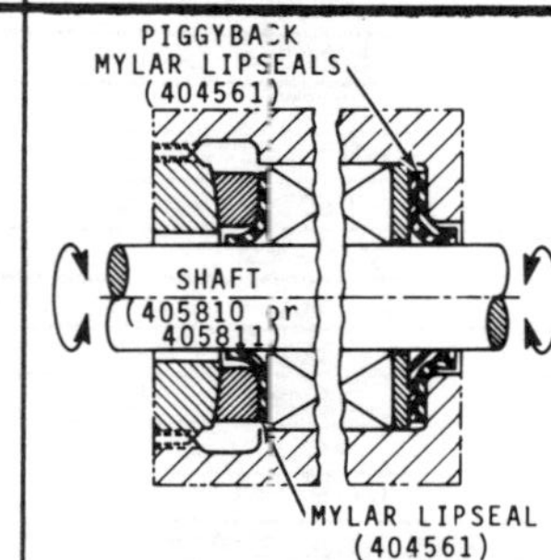	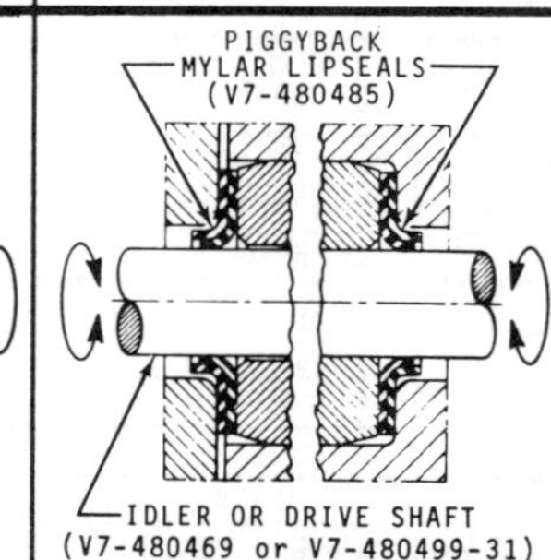
		SEAL TYPE	"U" – CUP FACE SEAL	LIPSEAL	LIPSEAL	LIPSEAL
		BASIC SEALING DIA., IN.	.975	.500	1.000	1.000
		DESIGN OPERATING MEDIA	GOX	GH2 or LH2	LH2	LOX or LH2
		DESIGN OPERATING PRESSURE, PSIG	1100	36	1130 psia	9 to 90
		DESIGN OPERATING TEMP. RANGE, °F	– 180 to + 860	– 423 to + 160	– 420 to + 140	– 423 to + 160
		MAXIMUM ALLOWABLE LEAKAGE	NOT SPECIFIED	25 SCCM	50 SCIM GHe @ 300 °F	50 SCIM GHe
		DESIGN OPERATING LIFE, CYCLES	2000	200	1308	750
		SPECIMEN OPERATION, CYCLES	147	30	292	591
INTERFACE MEMBERS DESCRIPTION	SEAL	PART NUMBER	5641026	511 – 51	404561	V7 – 480485
		PART NAME	SEAL – SHAFT	SEAL – MAIN SHAFT	SEAL, LIP – PROP. VALVE	LIPSEAL, 1 – INCH – FILL VALVE
		MATERIAL	AMS5667 INCONEL X	AMS3651 TEFLON	RB0130 – 039 MYLAR	RB0130 – 060 MYLAR
		SURFACE FINISH	8 r m s	SMOOTH MACHINE FINISH	AS REC'D PLASTIC FILM	AS REC'D PLASTIC FILM
		FORMING METHOD	MACHINED	8 r m s	COLD – FORMED	COLD – FORMED
		ADDITIONAL LOADING MEMBER (S)	NONE	511 – 52 SPRING – MAIN SHAFT SEAL	NONE; PIGGYBACK SEALS	PIGGYBACK SEALS
		BACK–UP SUPPORT MEMBER(S)	NONE	NONE	NONE	NONE
	MATING MEMBER	PART NUMBER(S)	5641525	511 – 55	405810, 405811	V7 – 480469, V7 – 480499 – 31
		PART NAME(S)	SHAFT – BUTTERFLY	SHAFT – MAIN	IDLER & DRIVE SHAFTS	IDLER & DRIVE SHAFTS
		MATERIAL	AMS5737 A – 286 CRES	AMS5667 INCONEL X – 750	AISI TYPE 440C CRES	MB0170 – 041 INCONEL 718
		SURFACE FINISH	8 r m s	8 r m s	8 r m s	8 r m s
		SURFACE TREATMENT	PASSIVATE	CHROME PLATE	NONE	THIN DENSE CHROME PLATE
		OTHER SURFACE TREATMENTS	NONE	PASSIVATE	NONE	PASSIVATE
USING COMPONENT DESCRIPTION		PROCUREMENT DOCUMENT(S) OR P/N(S)	ME284 – 0161 – 0008	1A49988	409920X7	V7 – 480450 – 121
		COMPONENT NAME	3 – INCH GOX PRESSURE REGULATOR	6 – INCH DIRECTIONAL CONTROL FUEL VENT VALVE	4 – INCH NORMALLY CLOSED MAIN FUEL VALVE	8 – INCH PROPELLANT FILL & DRAIN VALVE
		MANUFACTURER	PARKER – HANNIFIN	AMETEK / CALMEC	ROCKWELL / RKD	ROCKWELL / LAD
		MANUFACTURER'S PART NUMBER	5640012M7	511 – 509	409920X7	V7 – 480450 – 121
		USING VEHICLE(S) AND STAGE(S)	SATURN S – II	SATURN S – IVB	SATURN S – II & SATURN S – IVB	SATURN S – II
		USING SYSTEM(S)	LOX TANK PRESSURIZATION	LH_2 TANK VENT	J – 2 ENGINE LH_2 FEED	LOX & LH_2 TANK FILL & DRAIN

Interface Category Number 4 (Sheet 1 of 2)

ROTATING MAIN CLOSURE SEALS – Spherical or Face Sealing

		DESIGN CONCEPT NUMBER	4-1	4-2		
COMPONENT DATA BOOK		CONTROL NUMBER(S)	S – IVB / 210 – 3	S – II / 90 – 1	S – II / 100 – 1	S – II / 89 – 1
		INTERFACE DASH NUMBER(S)	– 12, & – 12A	– 1	– 1	– 1
		WEAR ANALYSIS RATING (AVG.)	**4.3**		**7.7***	
		PHOTOGRAPHS, PAGES	B – 8, – 15, – 16, & – 17	(6.9)* B – 2, – 8, & – 9	(8.0)* 81, 84, & 85	(8.2)* B – 2, & – 7
INTERFACE DESCRIPTION		INTERFACE PICTORIAL DESCRIPTION	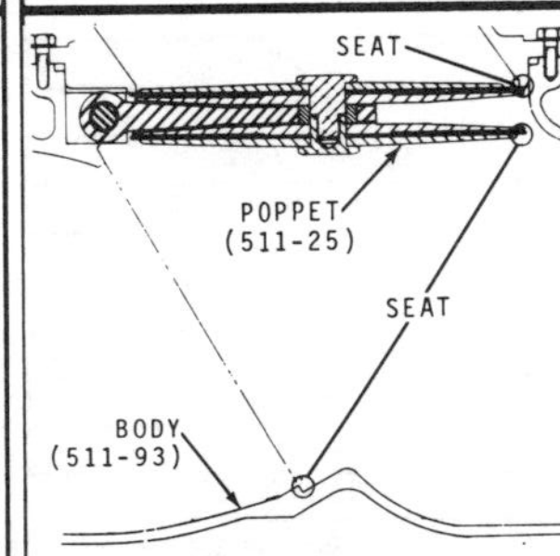	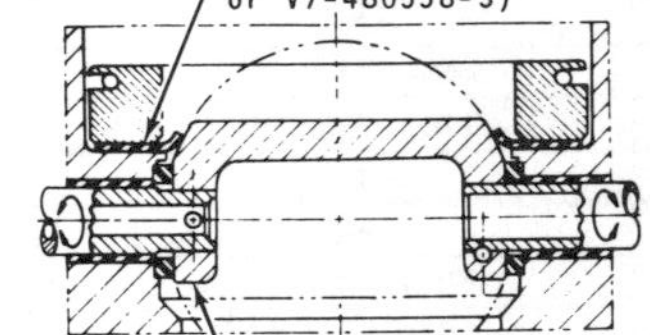		
		SEAL TYPE	FLAPPER POPPET SEAL	LIPSEAL	LIPSEAL	LIPSEAL
		BASIC SEALING DIA., IN.	6.162	8.035	1.820	8.050
		DESIGN OPERATING MEDIA	GH2 or LH2	LOX or LH2	LOX or LH2	LOX or LH2
		DESIGN OPERATING PRESSURE, PSIG	36	0 to 132	0 to 90	0 to 90
		DESIGN OPERATING TEMP. RANGE, °F	– 423 to + 160	– 423 to + 160	– 423 to + 160	– 423 to + 160
		MAXIMUM ALLOWABLE LEAKAGE	50 SCCM (3), 150 SCCM (D)	2500 SCIM GH_2 @ 50 PSIG & – 423 °F	500SCIM He @ 90 PSIG & – 320 °F	300SCIM He @ 90 PSIG & – 300 °F
		DESIGN OPERATING LIFE, CYCLES	200	1000	750	750
		SPECIMEN OPERATION, CYCLES	30	1178	935	591
INTERFACE MEMBERS DESCRIPTION	SEAL	PART NUMBER	511 – 25	5671452	V7 – 480558 – 3	V7 – 480743 – 3
		PART NAME	POPPET, MAIN	SEAL – BUTTERFLY	SEAL, GATE – RECIRC. VALVE	LIPSEAL, GATE – LOX PREVALVE
		MATERIAL	MIL – S – 5059 18 – 8 / 301 CRES	MB0130 – 053 KEL – F	MB0130 – 053 KEL – F	MB0130 – 053 KEL – F
		SURFACE FINISH	8 r m s	8 r m s	8 r m s	8 r m s
		FORMING METHOD	MACHINED	MACHINED	MACHINED	MACHINED
		ADDITIONAL LOADING MEMBER (S)	NONE	NONE	NONE	NONE
		BACK–UP SUPPORT MEMBER(S)	NONE	NONE	NONE	NONE
	MATING MEMBER	PART NUMBER	511 – 93	5671407 – 1	V7 – 480552	V7 – 480403 – 21
		PART NAME	BODY, VALVE 6" DIR. CONT. VV	BUTTERFLY – MACHINED	GATE – RECIRC. VALVE	GATE – FILL VALVE ASSY OF
		MATERIAL	QQ – A – 601 356 – T6 AL. ALY.	AMS5734 A – 286 CRES	AMS5737 A – 286 CRES	AMS5734 A – 286 CRES
		SURFACE FINISH	8 r m s	8 r m s	8 r m s	8 r m s
		SURFACE TREATMENT	MIL – A – 8625 ANODIZE	PASSIVATE	PASSIVATE	PASSIVATE
		OTHER SURFACE TREATMENTS	NONE	NONE	NONE	NONE
USING COMPONENT DESCRIPTION		PROCUREMENT DOCUMENT(S) OR P/N(S)	1A49988	ME284 – 0358 – 0001	V7 – 480328 – 91	V7 – 480450 – 121
		COMPONENT NAME	6 – INCH DIRECTIONAL CONTROL FUEL VENT VALVE	8 – INCH PROPELLANT SHUTOFF VALVE (PREVALVE)	2 – INCH NORMALLY OPEN RECIRCULATION VALVE	8 – INCH PROPELLANT FILL & DRAIN VALVE
		MANUFACTURER	AMETEK / CALMEC	PARKER – HANNIFIN	ROCKWELL / LAD	ROCKWELL / LAD
		MANUFACTURER'S PART NUMBER	511 – 509	5640024	V7 – 480328 – 91	V7 – 480450 – 121
		USING VEHICLE(S) AND STAGE(S)	SATURN S – IVB	SATURN S – II	SATURN S – II	SATURN S – II
		USING SYSTEM(S)	LH_2 TANK VENT	LOX & LH_2 FEED	LOX & LH_2 RECIRCULATION	LOX & LH_2 TANK FILL & DRAIN

Interface Category Number 4 (Sheet 2 of 2)

ROTATING MAIN CLOSURE SEALS - Spherical or Face Sealing

		4-3	4-4		
	DESIGN CONCEPT NUMBER	4-3	4-4		
COMPONENT DATA BOOK	CONTROL NUMBER(S)	S-IC/206-1	J-2/208-1		
COMPONENT DATA BOOK	INTERFACE DASH NUMBER(S)	-11, &-12	-1		
COMPONENT DATA BOOK	**WEAR ANALYSIS RATING (AVG.)**	**8.1**	**8.8**		
COMPONENT DATA BOOK	PHOTOGRAPHS, PAGES	B-2, -12 & -13	B-2, -3, -4, & -9		
INTERFACE DESCRIPTION	INTERFACE PICTORIAL DESCRIPTION	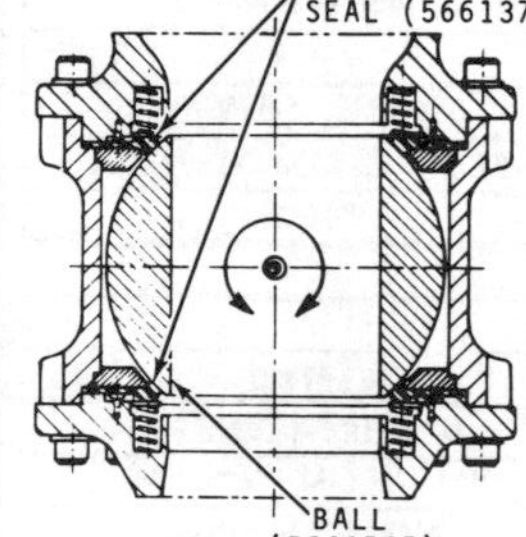	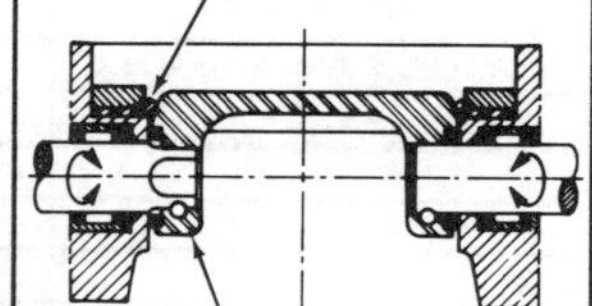		
INTERFACE DESCRIPTION	SEAL TYPE	SPHERICAL FACE SEAL	LIPSEAL		
INTERFACE DESCRIPTION	BASIC SEALING DIA., IN.	2.440	4.055		
INTERFACE DESCRIPTION	DESIGN OPERATING MEDIA	LH_2	LH_2		
INTERFACE DESCRIPTION	DESIGN OPERATING PRESSURE, PSIG	0 to 50	1130 PSIA		
INTERFACE DESCRIPTION	DESIGN OPERATING TEMP. RANGE, °F	-423 to +165	-420 to +140		
INTERFACE DESCRIPTION	MAXIMUM ALLOWABLE LEAKAGE	NO LIQUID (20 SCIM GH_2)	100 SCIM GHe @ 100 PSIG & -300°F		
INTERFACE DESCRIPTION	DESIGN OPERATING LIFE, CYCLES	2000	1308		
INTERFACE DESCRIPTION	SPECIMEN OPERATION, CYCLES	3318	292		
INTERFACE DESCRIPTION					
INTERFACE MEMBERS DESCRIPTION – SEAL	PART NUMBER	5661378 (2)	407089		
INTERFACE MEMBERS DESCRIPTION – SEAL	PART NAME	SEAL - BALL	SEAL, MAIN GATE		
INTERFACE MEMBERS DESCRIPTION – SEAL	MATERIAL	FLUORALOY "A" (FILLED TEFLON)	RB0130 - 005 KEL - F		
INTERFACE MEMBERS DESCRIPTION – SEAL	SURFACE FINISH	32 r m s	32 r m s		
INTERFACE MEMBERS DESCRIPTION – SEAL	FORMING METHOD	MACHINED	HOT - FORMED		
INTERFACE MEMBERS DESCRIPTION – SEAL	ADDITIONAL LOADING MEMBER (S)	5661675 SEAL SPRING	NONE		
INTERFACE MEMBERS DESCRIPTION – SEAL	BACK-UP SUPPORT MEMBER(S)	5641564 SEAL BACKUP	NONE		
INTERFACE MEMBERS DESCRIPTION – MATING MEMBER	PART NUMBER	5641535	405809		
INTERFACE MEMBERS DESCRIPTION – MATING MEMBER	PART NAME	BALL	GATE, MAIN FUEL VALVE		
INTERFACE MEMBERS DESCRIPTION – MATING MEMBER	MATERIAL	QQ - A - 225 / 8A 6061 - T6 AL. ALY.	(MAKE FROM 405917 ROUGH FORGING)		
INTERFACE MEMBERS DESCRIPTION – MATING MEMBER	SURFACE FINISH	8 r m s	8 r m s		
INTERFACE MEMBERS DESCRIPTION – MATING MEMBER	SURFACE TREATMENT	HARD ANODIZE	NONE		
INTERFACE MEMBERS DESCRIPTION – MATING MEMBER	OTHER SURFACE TREATMENTS	NONE	NONE		
INTERFACE MEMBERS DESCRIPTION – MATING MEMBER					
USING COMPONENT DESCRIPTION	PROCUREMENT DOCUMENT(S) OR P/N(S)	20M32014	409920X7		
USING COMPONENT DESCRIPTION	COMPONENT NAME	2 - INCH NORMALLY CLOSED SUBMERGED LH_2 SHUTOFF VALVE	4 - INCH NORMALLY CLOSED MAIN FUEL VALVE		
USING COMPONENT DESCRIPTION	MANUFACTURER	PARKER - HANNIFIN	ROCKWELL / RKD		
USING COMPONENT DESCRIPTION	MANUFACTURER'S PART NUMBER	5640027	409920X7		
USING COMPONENT DESCRIPTION	USING VEHICLE(S) AND STAGE(S)	SATURN S - IC (QUALIFIED BACKUP)	SATURN S - II & SATURN S - IVB		
USING COMPONENT DESCRIPTION	USING SYSTEM(S)	DNA	J - 2 ENGINE LH_2 FEED		

Interface Category Number 5 (Sheet 1 of 3)

RECIPROCATING POPPET, BALL, OR PISTON SEATS (PLASTIC) - Axial Seating

		DESIGN CONCEPT NUMBER	5-1	5-2	5-3	5-4
COMPONENT DATA BOOK		CONTROL NUMBER(S)	S – II / 96 – 5	S – II / 96 – 5	S – IVB / 210 – 3	S – II / 92 – 3
COMPONENT DATA BOOK		INTERFACE DASH NUMBER(S)	– 10	– 12	– 10 & – 10A	– 1
COMPONENT DATA BOOK		**WEAR ANALYSIS RATING (AVG.)**	**6.7**	**7.0**	**7.1**	**7.8**
COMPONENT DATA BOOK		PHOTOGRAPHS, PAGES	B – 7, – 11 & – 12	B – 4, – 14, & – 15	B – 3, – 4, & – 14	88
INTERFACE DESCRIPTION		INTERFACE PICTORIAL DESCRIPTION	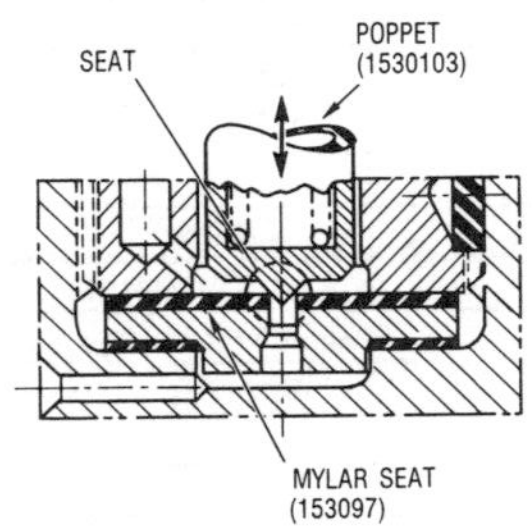	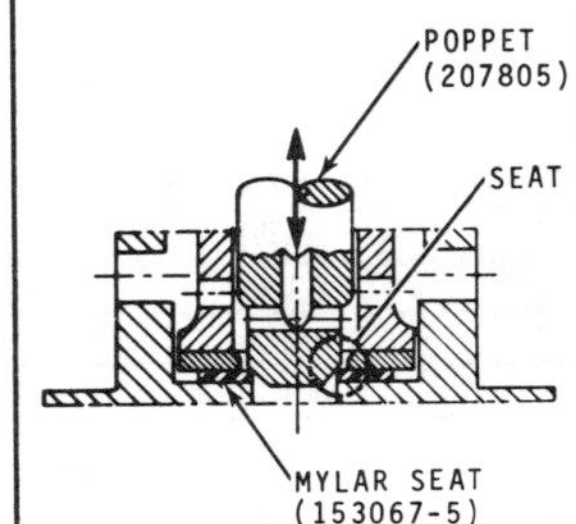	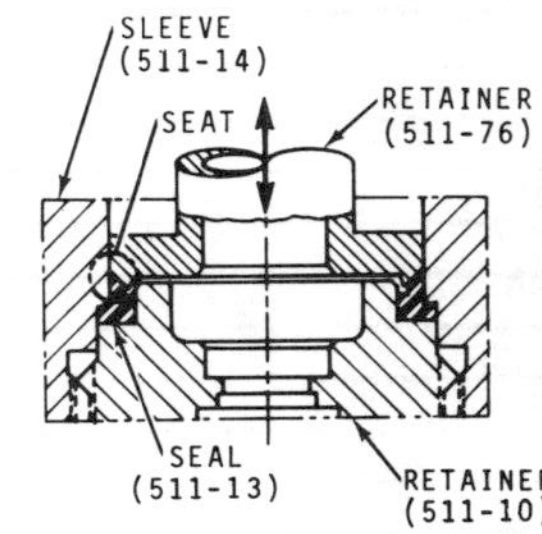	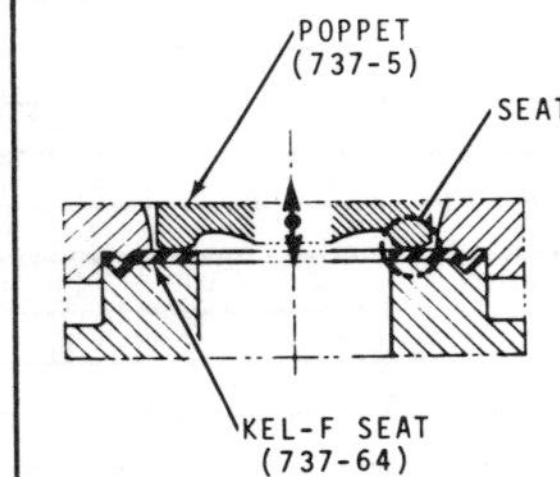
INTERFACE DESCRIPTION		SEAT TYPE	CONICAL	CONICAL	CONICAL	FLAT
INTERFACE DESCRIPTION		BASIC SEALING DIA., IN.	.018	.111	.805	7.000
INTERFACE DESCRIPTION		DESIGN OPERATING MEDIA	GHe	GHe	GHe	GHe & GH_2
INTERFACE DESCRIPTION		DESIGN OPERATING PRESSURE, PSIG	0 to 3500	850 to 3050	475	0 to 32.5
INTERFACE DESCRIPTION		DESIGN OPERATING TEMP. RANGE, °F	– 180 to + 160	– 180 to + 160	– 423 to + 160	– 423 to + 160
INTERFACE DESCRIPTION		MAXIMUM ALLOWABLE LEAKAGE	2000 SCIM GHe	50 SCIM GHe	50 SCCM	500 SCIM GHe; 2600 SCIM GH_2
INTERFACE DESCRIPTION		DESIGN OPERATING LIFE, CYCLES	150	500	200	600
INTERFACE DESCRIPTION		SPECIMEN OPERATION, CYCLES	195	565	30	1257
INTERFACE DESCRIPTION						
INTERFACE MEMBERS DESCRIPTION	SEAT	PART NUMBER	153097	153067 – 5	511 – 13	737 – 64
INTERFACE MEMBERS DESCRIPTION	SEAT	PART NAME	SEAT – RELIEF VALVE	SEAT – METERING ASSY, REGULATOR	SEAL, LOCK PISTON	SEAT – MAIN POPPET
INTERFACE MEMBERS DESCRIPTION	SEAT	MATERIAL	MIL – I – 631 TYPE A MYLAR	MIL – I – 631 TYPE A MYLAR	ANS3651 TEFLON	KEL – F
INTERFACE MEMBERS DESCRIPTION	SEAT	SURFACE FINISH	AS REC'D. PLASTIC FILM	AS REC'D. PLASTIC FILM	SMOOTH MACHINE FINISH (125 r m s)	AS REC'D. PLASTIC SHEET
INTERFACE MEMBERS DESCRIPTION	SEAT	RETENTION METHOD	CLAMPED	CLAMPED	CLAMPED	CLAMPED (SERRATION)
INTERFACE MEMBERS DESCRIPTION	SEAT					
INTERFACE MEMBERS DESCRIPTION	SEAT					
INTERFACE MEMBERS DESCRIPTION	MATING MEMBER	PART NUMBER	1530103	207805	511 – 76	737 – 5
INTERFACE MEMBERS DESCRIPTION	MATING MEMBER	PART NAME	POPPET – POPPET ASSY, R.V., REG	POPPET – METERING, REGULATOR	RETAINER, LOCK PISTON	POPPET – MAIN
INTERFACE MEMBERS DESCRIPTION	MATING MEMBER	MATERIAL	AMS5737 A – 286 CRES	AMS5737 A – 286 CRES	MIL – S – 7720 18 – 8 / 303 CRES	QQ – A – 282 7075 – T73 AL. ALY.
INTERFACE MEMBERS DESCRIPTION	MATING MEMBER	SURFACE FINISH	8 r m s	8 r m s	32 r m s	16 r m s
INTERFACE MEMBERS DESCRIPTION	MATING MEMBER	SURFACE TREATMENT	DRY FILM LUBE	DRY FILM LUBE	PASSIVATE	MIL – A – 8625 ANODIZE
INTERFACE MEMBERS DESCRIPTION	MATING MEMBER	OTHER SURFACE TREATMENTS	PASSIVATE	PASSIVATE	NONE	NONE
INTERFACE MEMBERS DESCRIPTION	MATING MEMBER					
USING COMPONENT DESCRIPTION		PROCUREMENT DOCUMENT(S) OR P/N(S)	ME284 – 0158 – 0004	ME284 – 0158 – 0004	1A49988	ME284 – 0030 – 0039
USING COMPONENT DESCRIPTION		COMPONENT NAME	3 / 8 – INCH HELIUM INJECTION REGULATOR – RELIEF VALVE	3 / 8 – INCH HELIUM INJECTION REGULATOR – RELIEF VALVE	6 – INCH DIRECTIONAL CONTROL FUEL VENT VALVE	7 – INCH LH_2 TANK VENT & RELIEF VALVE
USING COMPONENT DESCRIPTION		MANUFACTURER	ROYAL INDUSTRIES	ROYAL INDUSTRIES	AMETEK / CALMEC	AMETEK / CALMEC
USING COMPONENT DESCRIPTION		MANUFACTURER'S PART NUMBER	12078	12078	511 – 509	737 – 555
USING COMPONENT DESCRIPTION		USING VEHICLE(S) AND STAGE(S)	SATURN S – II	SATURN S – II	SATURN S – IVB	SATURN S – II
USING COMPONENT DESCRIPTION		USING SYSTEM(S)	VALVE ACTUATION, HELIUM INJECTION, & ACCUMULATOR FILL	VALVE ACTUATION, HELIUM INJECTION, & ACCUMULATOR FILL	LH_2 TANK VENT	LH_2 TANK VENT

Interface Category Number 5
(Sheet 2 of 3)

RECIPROCATING POPPET, BALL, OR PISTON SEATS
(PLASTIC) - Axial Seating

	DESIGN CONCEPT NUMBER	5-5	5-6	5-7	5-8
COMPONENT DATA BOOK	CONTROL NUMBER(S)	S - II / 92 - 3	S - IVB / 210 - 3	S - II / 92 - 3	S - II / 104 - 2; S - IC / I - 2
	INTERFACE DASH NUMBER(S)	- 12	- 1, - 1A, & - 1B	- 3	- 4 & - 5; - 3 & - 5
	WEAR ANALYSIS RATING (AVG.)	**8.1**	**8.2**	**8.2**	**8.3**
	PHOTOGRAPHS, PAGES	86	B - 2, - 3, - 4, - 5, & - 9	89, 94, & 95	B - 2, - 3, & - 7; B - 2, - 3, & - 5
INTERFACE DESCRIPTION	INTERFACE PICTORIAL DESCRIPTION	(see 5-5 drawing below)	(see 5-6 drawing below)	(see 5-7 drawing below)	(see 5-8 drawing below)

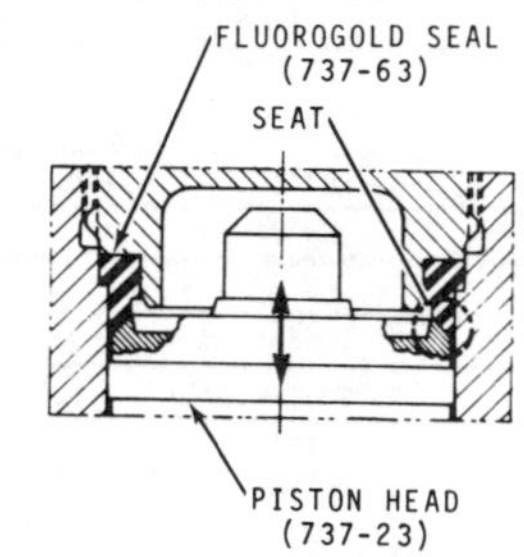

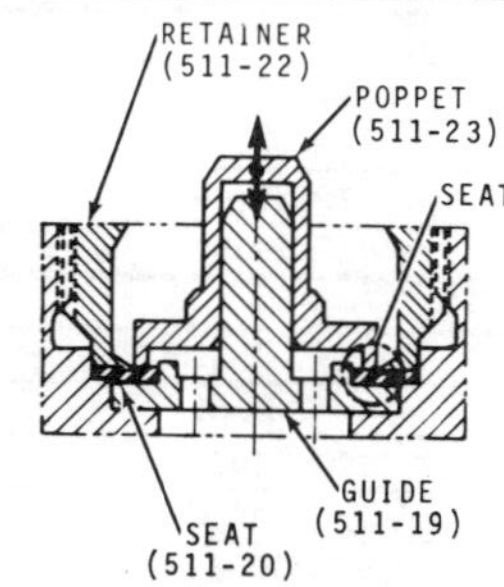

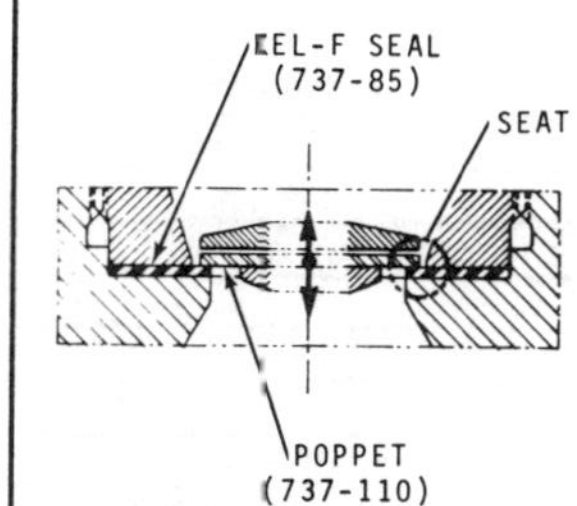

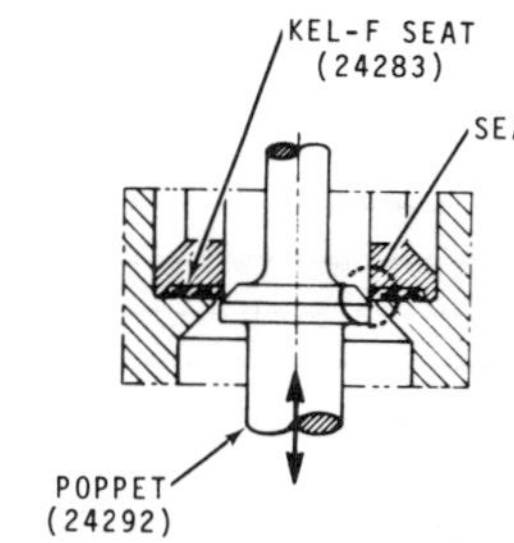

		5-5	5-6	5-7	5-8
INTERFACE DESCRIPTION	SEAT TYPE	CONICAL	FLAT	FLAT	CONICAL
	BASIC SEALING DIA., IN.	1.750	.312	1.375	.375
	DESIGN OPERATING MEDIA	GHe	GHe	GHe & GH_2	GHe
	DESIGN OPERATING PRESSURE, PSIG	0 to 750	LEAKAGE VENT ONLY	0 to 32.5	750
	DESIGN OPERATING TEMP. RANGE, °F	- 423 to + 160	- 423 to + 160	- 423 to + 160	- 180 to + 160
	MAXIMUM ALLOWABLE LEAKAGE	0.25 SCFM	NOT SPECIFIED	NO REQUIREMENT	45 SCIM
	DESIGN OPERATING LIFE, CYCLES	600	200	600	2000
	SPECIMEN OPERATION, CYCLES	1257	30	1257	1096; 5248
INTERFACE MEMBERS DESCRIPTION — SEAT	PART NUMBER	737 - 63	511 - 20	737 - 85	24283
	PART NAME	SEAL - VENT PISTON	SEAT, LOCK PISTON VENT C / V	SEAL - SECONDARY PILOT VALVE	SEAT, POPPET
	MATERIAL	FLUOROGOLD	AMS3651 TEFLON	KEL - F	KEL - F TEC GRADE B
	SURFACE FINISH	32 r m s	SMOOTH MACHINE FINISH	AS REC'D. PLASTIC SHEET	32 r m s
	RETENTION METHOD	CLAMPED	CLAMPED	CLAMPED	CLAMPED (SERRATION)
INTERFACE MEMBERS DESCRIPTION — MATING MEMBER	PART NUMBER	737 - 23	511 - 23	737 - 110	24292 & 24296
	PART NAME	HEAD - VENT ACTUATING PISTON	POPPET, VENT CHECK VALVE	POPPET - SECONDARY PILOT VALVE	POPPET, LOWER & UPPER
	MATERIAL	QQ - A - 225 / 8 6061 - T651 AL. ALY.	MIL - S - 7720 18 - 8 / 303 CRES	MIL - S - 5059A 18 - 8 / 301 CRES	QQ - S - 763 303 CRES
	SURFACE FINISH	32 r m s	16 r m s	6 r m s	8 r m s
	SURFACE TREATMENT	HARD ANODIZE	PASSIVATE	PASSIVATE	ELECTROPOLISH
	OTHER SURFACE TREATMENTS	NONE	NONE	NONE	NONE
USING COMPONENT DESCRIPTION	PROCUREMENT DOCUMENT(S) OR P/N(S)	ME284 - 0030 - 0039	1A49988	ME284 - 0030 - 0039	ME284 - 0356 - 0001; 60B52101 - 15F
	COMPONENT NAME	7 - INCH LH_2 TANK VENT & RELIEF VALVE	6 - INCH DIRECTIONAL CONTROL FUEL VENT VALVE	7 - INCH LH_2 TANK VENT & RELIEF VALVE	3 / 8 - INCH NORMALLY CLOSED SOLENOID VALVE
	MANUFACTURER	AMETEK / CALMEC	AMETEK / CALMEC	AMETEK / CALMEC	J. C. CARTER
	MANUFACTURER'S PART NUMBER	737 - 555	511 - 509	737 - 555	6969 - 2; 6969 - 15
	USING VEHICLE(S) AND STAGE(S)	SATURN S - II	SATURN S - IVB	SATURN S - II	SATURN S - II; SATURN S - IC
	USING SYSTEM(S)	LH_2 TANK VENT	LH_2 TANK VENT	LH_2 TANK VENT	VALVE ACTUATION, HELIUM INJECTION & ACCUMULAOTR FILL

Interface Category Number 5 (Sheet 3 of 3)

RECIPROCATING POPPET, BALL, OR PISTON SEATS (PLASTIC) - Axial Seating

		DESIGN CONCEPT NUMBER	5-9	5-10	5-11	
COMPONENT DATA BOOK		CONTROL NUMBER(S)	S – II / 100 – 1	S – IVB / 209 – 2	S – IVB / 209 – 2	
		INTERFACE DASH NUMBER(S)	– 11	– 8	– 14	
		WEAR ANALYSIS RATING (AVG.)	**8.3**	**9.4**	**9.4**	
		PHOTOGRAPHS, PAGES	88, 94, & 95	B – 5, & – 15	B – 6, & – 18	
INTERFACE DESCRIPTION		INTERFACE PICTORIAL DESCRIPTION	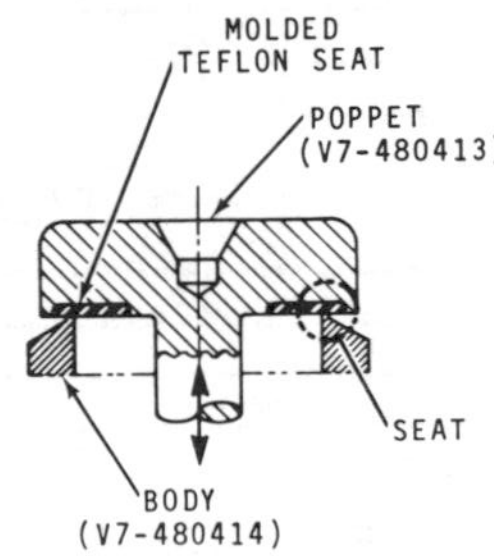	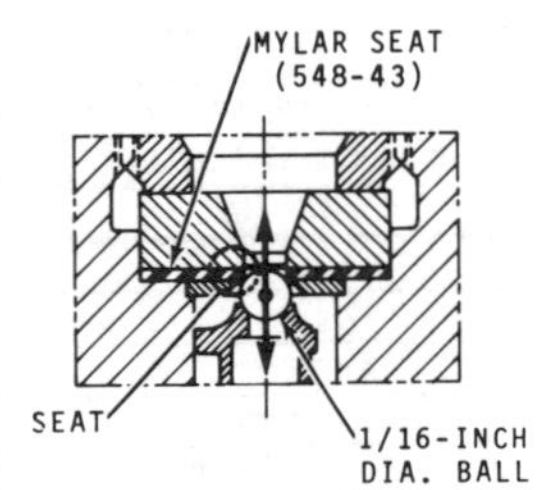	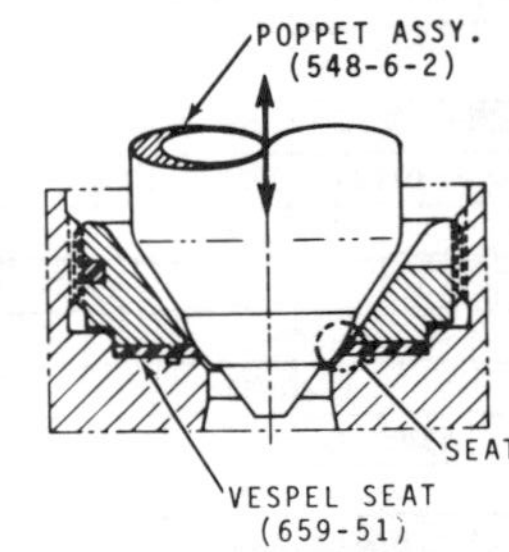	
		SEAT TYPE	FLAT	SPHERICAL	CONICAL	
		BASIC SEALING DIA., IN.	.320	.053	.379	
		DESIGN OPERATING MEDIA	LOX or LH_2	GHe	GHe	
		DESIGN OPERATING PRESSURE, PSIG	45 + 10, – 5 (CRACKING PRESSURE)	0 to 1000	25 to 3100	
		DESIGN OPERATING TEMP. RANGE, °F	– 423 to + 160	– 423 to + 160	– 423 to + 160	
		MAXIMUM ALLOWABLE LEAKAGE	NOT ISOLATED	NOT SPECIFIED	100 SCIM @ 400 PSIG, & *	
		DESIGN OPERATING LIFE, CYCLES	750	5000	5000	
		SPECIMEN OPERATION, CYCLES	935	10,121	10,121	
					* 40 SCIM @ 3000 PSIG	
INTERFACE MEMBERS DESCRIPTION	SEAT	PART NUMBER	V7 – 480413	548 – 43	659 – 51	
		PART NAME	POPPET – VALVE, RELIEF	SEAT, PILOT VALVE	SEAT, SHUTOFF VALVE POPPET	
		MATERIAL	MB0130 – 052 TEFLON	MYLAR "D"	STM0494 VESPEL	
		SURFACE FINISH	16 r m s	SHARP AND SMOOTH	16 r m s	
		RETENTION METHOD	MOLDED RING SEAT	CLAMPED	CLAMPED	
	MATING MEMBER	PART NUMBER	V7 – 480414	NONE	548 – 38	
		PART NAME	BODY – VALVE, RELIEF, ASSY OF	1 / 16 DIA. BALL	POPPET, SOLENOID VALVE	
		MATERIAL	QQ – S – 763 303Se CRES	440C CRES	AMS5643 17 – 4 CRES	
		SURFACE FINISH	16 r m s	NOT SPECIFIED	4 r m s	
		SURFACE TREATMENT	NONE	NONE	DENSE CHROME PLATE	
		OTHER SURFACE TREATMENTS	NONE	NONE	PASSIVATE	
USING COMPONENT DESCRIPTION		PROCUREMENT DOCUMENT(S) OR P/N(S)	V7 – 480328 – 91	1B43660 – 515	1B43660 – 515	
		COMPONENT NAME	2 – INCH NORMALLY OPEN RECIRCULATION VALVE	1 / 2 – INCH NORMALLY CLOSED HELIUM SOLENOID VALVE	1 / 2 – INCH NORMALLY CLOSED HELIUM SOLENOID VALVE	
		MANUFACTURER	ROCKWELL / LAD	AMETEK / CALMEC	AMETEK / CALMEC	
		MANUFACTURER'S PART NUMBER	V7 – 480328 – 91	548 – 515	548 – 515	
		USING VEHICLE(S) AND STAGE(S)	SATURN S – II	SATURN S – IVB	SATURN S – IVB	
		USING SYSTEM(S)	LOX & LH_2 RECIRCULATION	PRESSURE CONTROL	PRESSURE CONTROL	

Interface Category Number 6
(Sheet 1 of 3)

RECIPROCATING POPPET OR BALL SEATS
(METAL-TO-METAL) - Axial Seating

	DESIGN CONCEPT NUMBER	6-1	6-2	6-3	6-4
COMPONENT DATA BOOK	CONTROL NUMBER(S)	S – IC / 207 – 5	S – II / 90 – 1	S – II / 98 – 5	S – IC / 207 – 5
	INTERFACE DASH NUMBER(S)	– 20	– 2	– 1	– 16, – 17, – 18, & – 19
	WEAR ANALYSIS RATING (AVG.)	**6.9**	**7.1**	**7.2**	**7.3**
	PHOTOGRAPHS, PAGES	B – 9, & – 21	B – 2, & – 10	B – 7, – 9, & – 10	B – 8, & – 20
INTERFACE DESCRIPTION	INTERFACE PICTORIAL DESCRIPTION	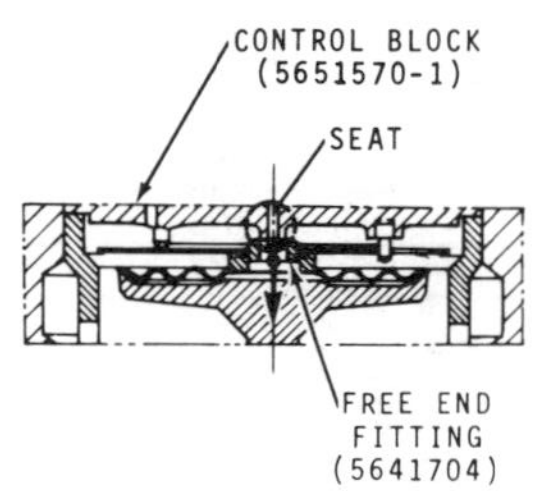	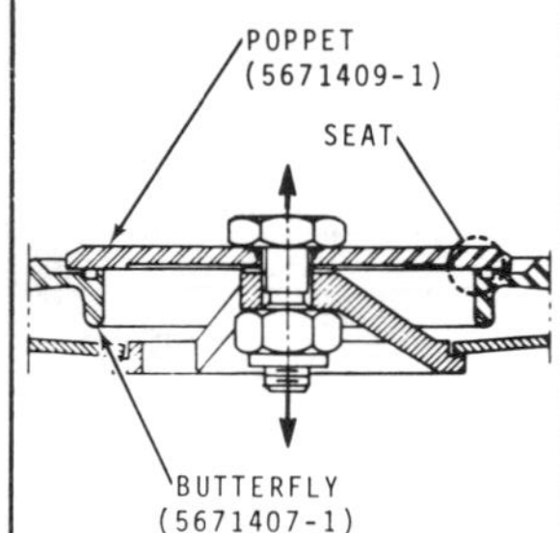	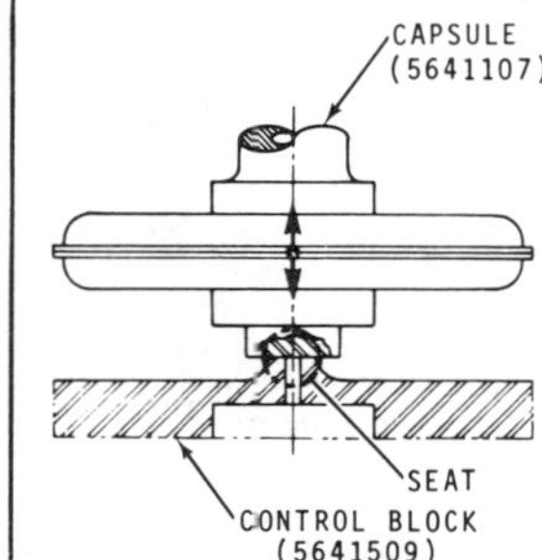	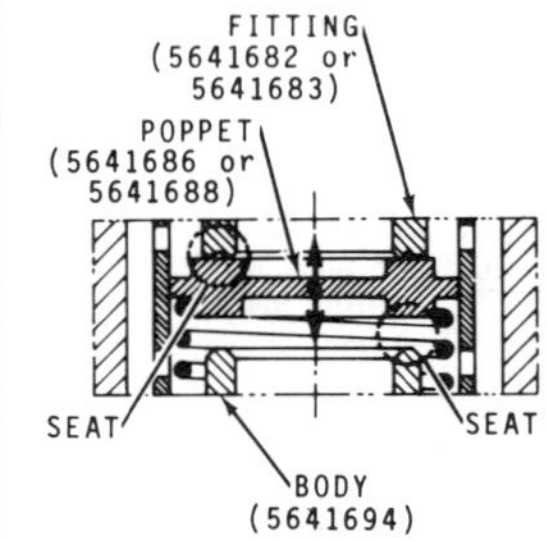
	SEAT TYPE	FLAT	FLAT	FLAT	FLAT
	BASIC SEALING DIA., IN.	.043	2.000	.035	.225 to .470
	DESIGN OPERATING MEDIA	GHe, GH_2 or RP – 1	LOX or LH_2	GOX	GHe, GH_2 or RP – 1
	DESIGN OPERATING PRESSURE, PSIG	to 35	0 to 50	36 to 37.5 PSIA	to 35
	DESIGN OPERATING TEMP. RANGE, °F	– 400 to + 165	– 423 to + 160	– 180 to + 860	– 400 to + 165
	MAXIMUM ALLOWABLE LEAKAGE	NOT SPECIFIED	1500 SCIM (INCLUDES GATE SEAL)	NOT SPECIFIED	NOT SPECIFIED
	DESIGN OPERATING LIFE, CYCLES	500	150 (1000)	2000	500
	SPECIMEN OPERATION, CYCLES	500	(1178)	147	500
INTERFACE MEMBERS DESCRIPTION — SEAT	PART NUMBER(S)	5651570 – 1 (5641680)	5671407 – 1	5641509	5641682 & 5641683; 5641694
	PART NAME	CONTROL BLOCK	BUTTERFLY – MACHINED	CONTROL BLOCK	FITTINGS; BODY
	MATERIAL	AMS5737 A – 286 CRES (STERCON H)	AMS5734 A – 286 CRES	MIL – S – 7720 303Se CRES	A – 286 CRES; 7075 – T73 AL. ALY.
	HEAT TREATMENT	AS REC'D. MAT'L. COND.	R_c 29 MINIMUM	COND. A	AS REC'D MAT'L COND.; –T73 COND.
	SURFACE FINISH	16 r m s	4 r m s	16 r m s	16 r m s; 125 r m s
	SURFACE TREATMENT	PASSIVATE	PASSIVATE	PASSIVATE	PASSIVATE; ANODIZE
	OTHER SURFACE TREATMENTS	NONE	NONE	NONE	NONE
INTERFACE MEMBERS DESCRIPTION — MATING MEMBER	PART NUMBER(S)	5641704	5671409 – 1	5641107	5641686; 5641688
	PART NAME(S)	FREE END FITTING, P.V. CAPSULE	POPPET – RELIEF VALVE	CAPSULE – NON – NESTED	POPPET, TANK; POPPET, OUTLET
	MATERIAL	QQ – A – 763 321 or 347 CRES	AMS5664 INCONEL 718	QQ – S – 763 347 CRES	AMS737 A – 286 CRES (STERCON H)
	HEAT TREATMENT	COND. A	R_C 38 MINIMUM	COND. A	AS REC'D. MAT'L. COND.
	SURFACE FINISH	16 r m s	4 r m s	16 r m s	16 r m s
	SURFACE TREATMENT	PASSIVATE	PASSIVATE	PASSIVATE	PASSIVATE
	OTHER SURFACE TREATMENTS	NONE	NONE	NONE	NONE
USING COMPONENT DESCRIPTION	PROCUREMENT DOCUMENT(S) OR P/N(S)	20M32015	ME284 – 0358 – 0001	ME284 – 0161 – 0008	20M32015
	COMPONENT NAME	1 – 1 / 2 INCH HELIUM PRESSURE REGULATOR	8 – INCH PROPELLANT SHUTOFF VALVE (PREVALVE)	3 – INCH GOX PRESSURE REGULATOR	1 – 1 / 2 INCH HELIUM PRESSURE REGULATOR
	MANUFACTURER	PARKER – HANNIFIN	PARKER – HANNIFIN	PARKER – HANNIFIN	PARKER – HANNIFIN
	MANUFACTURER'S PART NUMBER	5640061	5670024	5640012M7	5640061
	USING VEHICLE(S) AND STAGE(S)	SATURN S – IC (QUALIFIED BACK – UP)	SATURN S – II	SATURN S – II	SATURN S – IC (QUALIFIED BACK – UP)
	USING SYSTEM(S)	DNA	PROPELLANT FEED	PREVIOUSLY IN LOX TANK PRESSURIZATION	DNA

Interface Category Number 6 (Sheet 2 of 3)

RECIPROCATING POPPET OR BALL SEATS (METAL-TO-METAL) - Axial Seating

		DESIGN CONCEPT NUMBER	6-5	6-6	6-7	6-8
COMPONENT DATA BOOK		CONTROL NUMBER(S)	S-IC/207-5	S-II/98-5	S-II/98-5	S-IC/207-5
		INTERFACE DASH NUMBER(S)	-21	-2	-9, -10, -11, & -12	-1
		WEAR ANALYSIS RATING (AVG.)	**7.5**	**7.6**	**7.6**	**7.9**
		PHOTOGRAPHS, PAGES	B-10, & -22	B-5, & -11	B-6, -7, -14, -15, -16, -17, & -18	B-3, -5, -6, -7, -11, & -12
INTERFACE DESCRIPTION		INTERFACE PICTORIAL DESCRIPTION	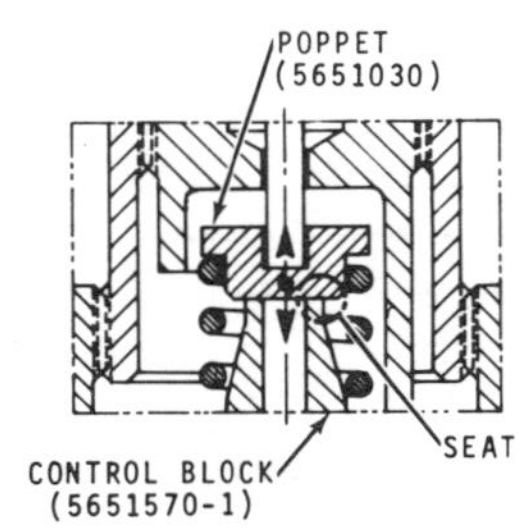	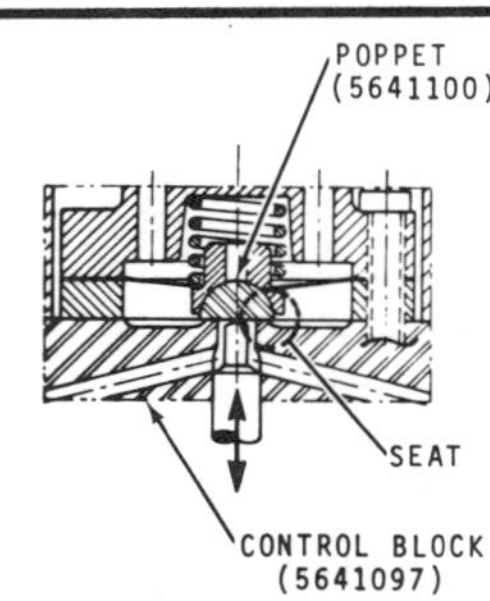	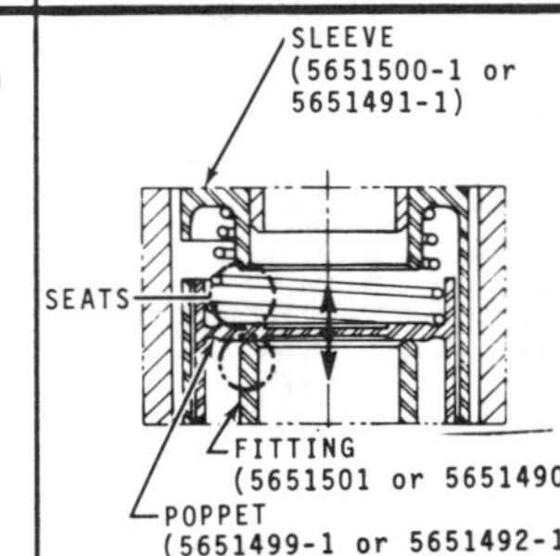	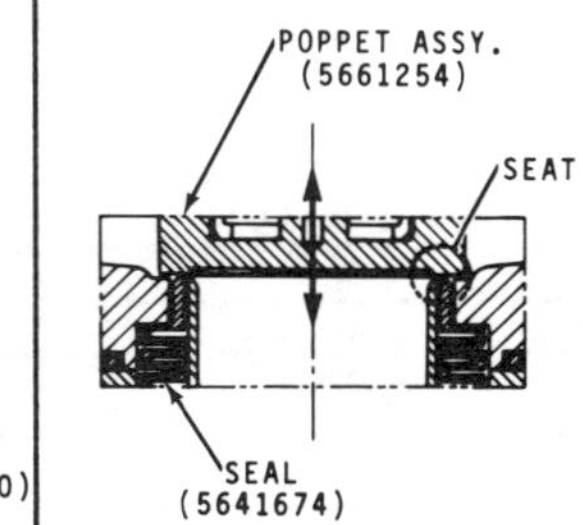
		SEAT TYPE	FLAT	FLAT	FLAT	FLAT
		BASIC SEALING DIA., IN.	.055	.097	.186 to .470	1.300
		DESIGN OPERATING MEDIA	GHe, GN_2 or RP-1	GOX	GOX	GHe, or GN_2
		DESIGN OPERATING PRESSURE, PSIG	to 35	APPROX. 500 PSI	0 to 1100	3200
		DESIGN OPERATING TEMP. RANGE, °F	-400 to +165	-180 to +860	-180 to +860	-400 to +165
		MAXIMUM ALLOWABLE LEAKAGE	NOT SPECIFIED	NOT SPECIFIED	(-9) 25 SCIM He; (-10) 30 SCIM GN_2 *	50 SCIM
		DESIGN OPERATING LIFE, CYCLES	500	2000	2000	500
		SPECIMEN OPERATION, CYCLES	500	147	147	500
					*(-11) 50 SCIM He; (-12) 30 SCIM GN_2	
INTERFACE MEMBERS DESCRIPTION	SEAT	PART NUMBER(S)	5651570-1 (5641680)	5641097	5651490 & 1; 5651500-1 & 1	5641647
		PART NAME(S)	CONTROL BLOCK - PILOT VALVE	CONTROL BLOCK - BIAS REG.	FITTINGS & SLEEVES	SEAL, POPPET
		MATERIAL	AMS5737 A-286 CRES (STERCON H)	MIL-S-7720 303 Se CRES	AMS5737 A-286 CRES (STERCON H)	AMS5667 INCONEL X750
		HEAT TREATMENT	AS REC'D. MAT'L. COND.	COND. B	AS REC'D, MAT'L. COND.	AS REC'D, MAT'L. COND
		SURFACE FINISH	8 rms	10 rms	2 rms (FITTINGS); & 4 rms (SLEEVES)	2 rms
		SURFACE TREATMENT	PASSIVATE	PASSIVATE	PASSIVATE	PASSIVATE
		OTHER SURFACE TREATMENTS	NONE	NONE	NONE	(OPT. SILVER, GOLD or TFE PLTG)
	MATING MEMBER	PART NUMBER(S)	5651030	5641100	5651499-1, 5651492-1	5661254-1
		PART NAME	POPPET - FLOW LIMITER	POPPET - BIAS REGULATOR	POPPETS	POPPET
		MATERIAL	AMS5737 A-286 CRES (STERCON H)	HAYNES STELLITE # 3	AMS5737 A-286 CRES (STERCON H)	AMS5737 A-286 CRES (STERCON H)
		HEAT TREATMENT	AS REC'D. MAT'L. COND.	AS REC'D. MAT'L. COND.	AS REC'D. MAT'L. COND.	AS REC'D. MAT'L. COND.
		SURFACE FINISH	2 rms	8 rms	4 rms	2 rms
		SURFACE TREATMENT	PASSIVATE	NONE	PASSIVATE	MICROPLATE NO. 1
		OTHER SURFACE TREATMENTS	NONE	NONE	NONE	PASSIVATE
USING COMPONENT DESCRIPTION		PROCUREMENT DOCUMENT(S) OR P/N(S)	20M32015	ME284-0161-0008	ME284-0161-0008	20M32015
		COMPONENT NAME	1-1/2 INCH HELIUM PRESSURE REGULATOR	3-INCH GOX PRESSURE REGULATOR	3-INCH GOX PRESSURE REGULATOR	1-1/2 INCH HELIUM PRESSURE REGULATOR
		MANUFACTURER	PARKER-HANNIFIN	PARKER-HANNIFIN	PARKER-HANNIFIN	PARKER-HANNIFIN
		MANUFACTURER'S PART NUMBER	5640061	5640012M7	5640012M7	5640061
		USING VEHICLE(S) AND STAGE(S)	SATURN S-IC (QUALIFIED BACK-UP)	SATURN S-II	SATURN S-II	SATURN S-IC QUALIFIED BACK-UP)
		USING SYSTEM(S)	DNA	PREVIOUSLY IN LOX TANK PRESSURIZATION	PREVIOUSLY IN LOX TANK PRESSURIZATION	DNA

Interface Category Number 6
(Sheet 3 of 3)

RECIPROCATING POPPET OR BALL SEATS
(METAL–TO–METAL) – Axial Seating

		DESIGN CONCEPT NUMBER	6-9	6-10	6-11	6-12
COMPONENT DATA BOOK		CONTROL NUMBER(S)	J – 2 / 208 – 1	S – II / 98 – 5	S – II / 98 – 5	S – II / 92 – 3
		INTERFACE DASH NUMBER(S)	– 18	– 8	– 18	– 6
		WEAR ANALYSIS RATING (AVG.)	**7.9**	**8.1**	**8.5**	**8.7**
		PHOTOGRAPHS, PAGES	B – 7, & – 22	B – 2, & – 7	B – 3, & – 21	85
INTERFACE DESCRIPTION		INTERFACE PICTORIAL DESCRIPTION	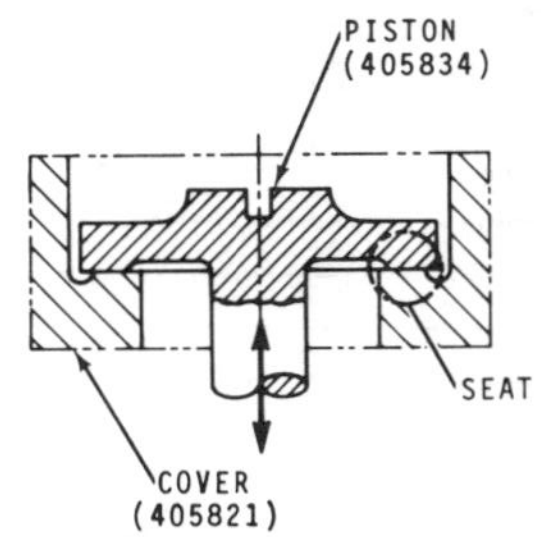	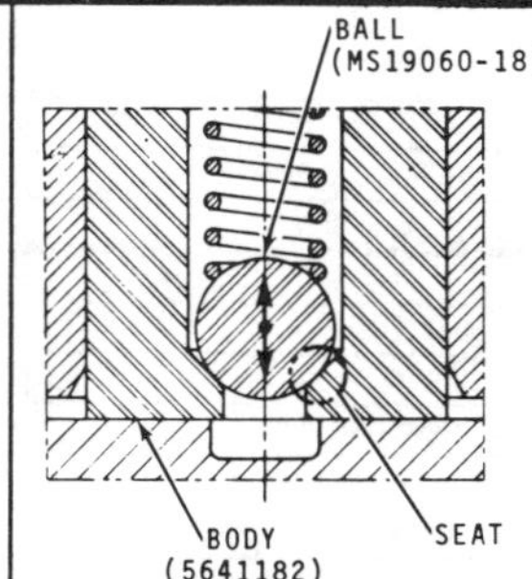	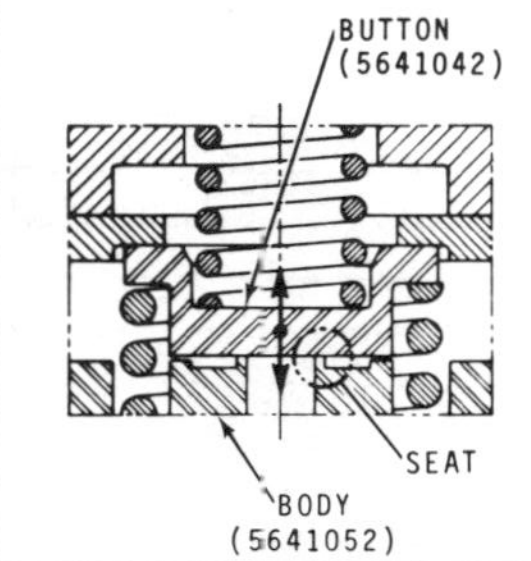	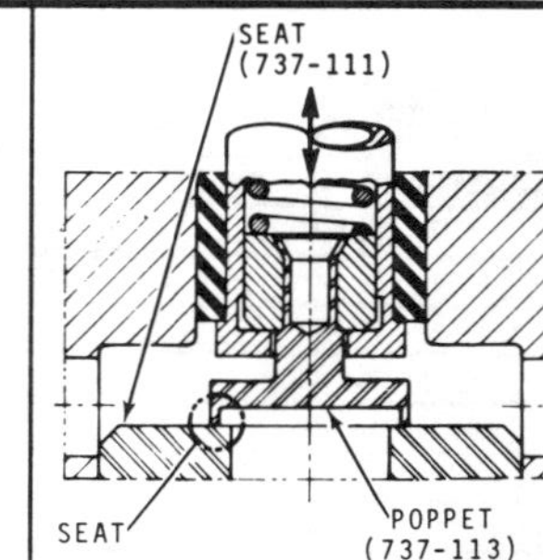
		SEAT TYPE	FLAT	CONICAL – SPHERICAL	FLAT	FLAT
		BASIC SEALING DIA., IN.	.875	.073	.093	.187
		DESIGN OPERATING MEDIA	GHe	GOX	GOX	GHe or GH_2
		DESIGN OPERATING PRESSURE, PSIG	400	10 PSID	475 – 525	0 to 32.5
		DESIGN OPERATING TEMP. RANGE, °F	– 420 to + 140	– 180 to + 860	– 180 to + 860	– 423 to + 160
		MAXIMUM ALLOWABLE LEAKAGE	300 SCIM GHe, AMB; 500 SCIM GHe, L.T.	NOT SPECIFIED	2.75 SCIM GN_2 @ 450 PSIA (AMB.)	NO REQUIREMENT
		DESIGN OPERATING LIFE, CYCLES	1308	2000	2000	600
		SPECIMEN OPERATION, CYCLES	292	147	147	1257
INTERFACE MEMBERS DESCRIPTION	SEAT	PART NUMBER	405821	5641182	5641052	737 – 111
		PART NAME	COVER, ACTUATOR, MFV	BODY – CHECK VALVE	BODY – MACHINED	SEAT – PRIMARY PILOT POPPET
		MATERIAL	RB0170 – 011 TENS 50 – T6 AL. ALY.	MIL – S – 7720 303 Se CRES	AMS5350 410 CRES	QQ – S – 763 440C CRES
		HEAT TREATMENT	COND. – T6	COND. B	INDUCTION HARDENED	R_C 58 – 60
		SURFACE FINISH	32 r m s	63 r m s	8 r m s	4 r m s
		SURFACE TREATMENT	CHROMIC ACID ANODIZE	PASSIVATE	PASSIVATE	PASSIVATE
		OTHER SURFACE TREATMENTS	NONE	NONE	NONE	NONE
	MATING MEMBER	PART NUMBER	405834	MS19060 – 18	5641042	737 – 113
		PART NAME	PISTON, SEQUENCE VALVE	BALL – BALL CHECK VALVE	BUTTON, PLUNGER ASSY.	POPPET – PRIMARY PILOT
		MATERIAL	AMS5735 A – 286 CRES	FED. STD. 66 440A, 440B or 440C CRES	AMS5737 A – 286 CRES (STERCON H)	AMS5643 17 – 4 PH CRES
		HEAT TREATMENT	AS REC'D. MAT'L. COND.	R_C 58 – 65	AS REC'D. MAT'L. COND.	COND. H – 925
		SURFACE FINISH	4 r m s	3 MICROINCHES MAX. SUR. ROUGH	4 r m s	4 r m s
		SURFACE TREATMENT	NONE	PASSIVATE	PASSIVATE	PASSIVATE
		OTHER SURFACE TREATMENTS	NONE	NONE	NONE	NONE
USING COMPONENT DESCRIPTION		PROCUREMENT DOCUMENT(S) OR P/N(S)	409920X7	ME284 – 0161 – 0008	ME284 – 0161 – 0008	ME284 – 0030 – 0039
		COMPONENT NAME	4 – INCH NORMALLY CLOSED MAIN FUEL VALVE	3 – INCH GOX PRESSURE REGULATOR	3 – INCH GOX PRESSURE REGULATOR	7 – INCH LH_2 TANK VENT & RELIEF VALVE
		MANUFACTURER	ROCKWELL / RKD	PARKER – HANNIFIN	PARKER – HANNIFIN	AMETEK / CALMEC
		MANUFACTURER'S PART NUMBER	409920X7	5640012M7	5640012M7	737 – 555
		USING VEHICLE(S) AND STAGE(S)	SATURN S – II & SATURN S – IVB	SATURN S – II	SATURN S – II	SATURN S – II
		USING SYSTEM(S)	J – 2 ENGINE LH_2 FEED	PREVIOUSLY IN LOX TANK PRESSURIZATION	PREVIOUSLY IN LOX TANK PRESSURIZATION	LH_2 TANK VENT

Interface Category Number 7
(Sheet 1 of 2)

RECIPROCATING SHAFT OR PISTON GUIDES (PLASTIC) - Radial Bearing

		DESIGN CONCEPT NUMBER	7-1	7-2	7-3	7-4
COMPONENT DATA BOOK		CONTROL NUMBER(S)	S-II/96-5	S-II/92-3	S-II/100-1	S-II/89-1
		INTERFACE DASH NUMBER(S)	-6	-7 & -7A	-8	-12
		WEAR ANALYSIS RATING (AVG.)	**5.7**	**8.6**	**8.8**	**8.8**
		PHOTOGRAPHS, PAGES	B-4, -9, & -10	87, 88, 96, & 97	NONE	B-3, -4, -5, -6, -10, & -12
INTERFACE DESCRIPTION		INTERFACE PICTORIAL DESCRIPTION	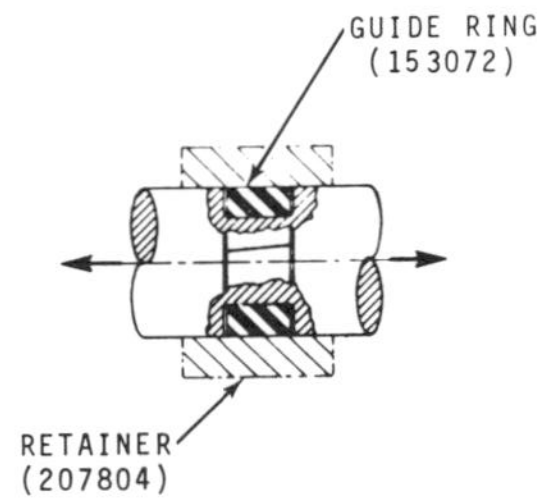	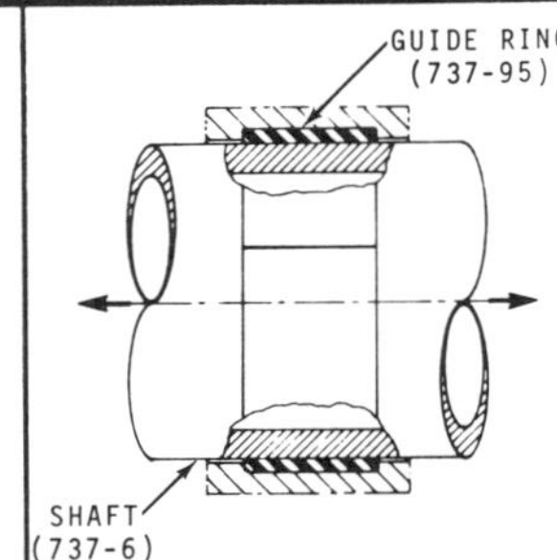	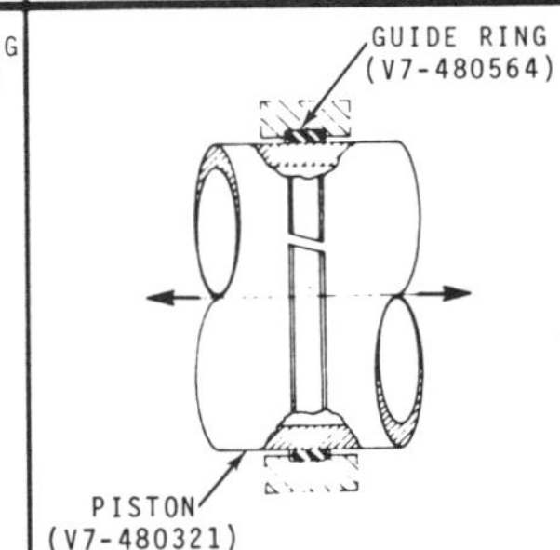	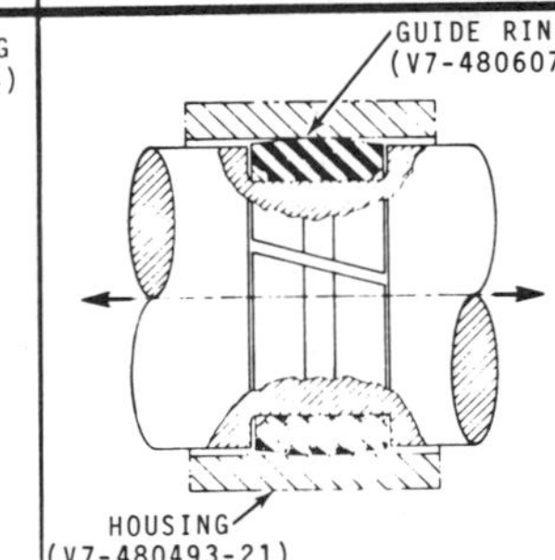
		GUIDE TYPE	SPLIT - RING	SPLIT - RING	SPLIT - RING	SPLIT - RING
		BASIC GUIDE BEARING DIA., IN.	.128	1.7490	1.750	.984
		GUIDE WIDTH, IN.	.051	.715	.185	.080
		L/D	.398	.410	.105	.081
		ASSOCIATED GUIDE MEMBERS	NONE	ONE OF TWO GUIDES	ONE OF THREE GUIDES	ONE OF TWO GUIDES
		DESIGN OPERATING MEDIA	GHe	GHe or GH_2	GHe	AIR
		DESIGN OPERATING TEMP. RANGE, °F	-180 to +160	-423 to +160	-423 to +160	-423 to +160
		DESIGN OPERATING LIFE, CYCLES	500	600	750	750
		SPECIMEN OPERATION, CYCLES	565	1257	935	591
INTERFACE MEMBERS DESCRIPTION	GUIDE	PART NUMBER	153072	737-95	V7-480564	V7-480607
		PART NAME	SEAL - PIN, METERING, REGULATOR	BEARING - SHAFT MAIN	BEARING, PISTON - RECIRC. VALVE	BEARING, PISTON SHAFT - PREVALVE
		MATERIAL	AMS3651 TEFLON	FLUOROGOLD	MB0130-052 TEFLON	MB0130-052 TEFLON
		SURFACE FINISH	16 r m s	SMOOTH MACHINE FINISH	125 r m s	125 r m s
		THICKNESS	.026	.060	.048	.116
	MATING MEMBER	PART NUMBER	207804	737-6	V7-480321	V7-480493-21
		PART NAME	RETAINER ASSY - METERING, REG.	SHAFT - MAIN POPPET	PISTON, RECIRC. VALVE - N. O.	HOUSING - FILL VALVE, ASSY OF
		MATERIAL	QQ-A-268 2024-T351 AL. ALY.	MIL-S-7720 18-8/303 CRES	AMS5737 A-286 CRES	MB170-029 TENS 50-T6 AL. ALY.
		SURFACE FINISH	16 r m s	8 r m s	8 r m s	32 r m s
		SURFACE TREATMENT	MIL-A-8625 ANODIZE	HARD CHROME PLATE	THIN DENSE CHROME PLATE	MA0109-009 ANODIZE
		OTHER SURFACE TREATMENTS	NONE	PASSIVATE	PASSIVATE	NONE
USING COMPONENT DESCRIPTION		PROCUREMENT DOCUMENT(S) OR P/N(S)	ME284-1058-0004	ME284-0030-0039	V7-480328-91	V7-480450-121
		COMPONENT NAME	3/8-INCH HELIUM INJECTION REGULATOR - RELIEF VALVE	7-INCH LH_2 TANK VENT & RELIEF VALVE	2-INCH NORMALLY OPEN RECIRCULATION VALVE	8-INCH PROPELLANT FILL & DRAIN VALVE
		MANUFACTURER	ROYAL INDUSTRIES	AMETEK/CALMEC	ROCKWELL/LAD	ROCKWELL/LAD
		MANUFACTURER'S PART NUMBER	12078	737-555	V7-480328-91	V7-480450-121
		USING VEHICLE(S) AND STAGE(S)	SATURN S-II	SATURN S-II	SATURN S-II	SATURN S-II
		USING SYSTEM(S	VALVE ACTUATION, HELIUM INJECTION, & ACCUMULATOR FILL	LH_2 TANK VENT	LOX & LH_2 RECIRCULATION	LOX & LH_2 TANK FILL & DRAIN

Interface Category Number 7
(Sheet 2 of 2)

RECIPROCATING SHAFT OR PISTON GUIDES
(PLASTIC) - Radial Bearing

Section		DESIGN CONCEPT NUMBER	7-5	7-6	7-7	7-8
COMPONENT DATA BOOK		CONTROL NUMBER(S)	S-II/100-1	S-IC/206-1	S-II/89-1	S-II/92-3
		INTERFACE DASH NUMBER(S)	-6 & -6A	-8 & -10	-15	-5
		WEAR ANALYSIS RATING (AVG.)	**9.2**	**9.4**	**9.5**	**9.8**
		PHOTOGRAPHS, PAGES	82, & 88	B-4, -10, & -11	B-5, -16, & -17	85
INTERFACE DESCRIPTION		INTERFACE PICTORIAL DESCRIPTION	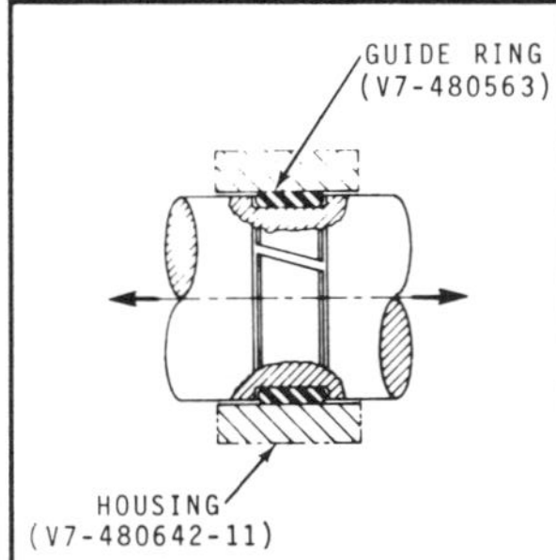	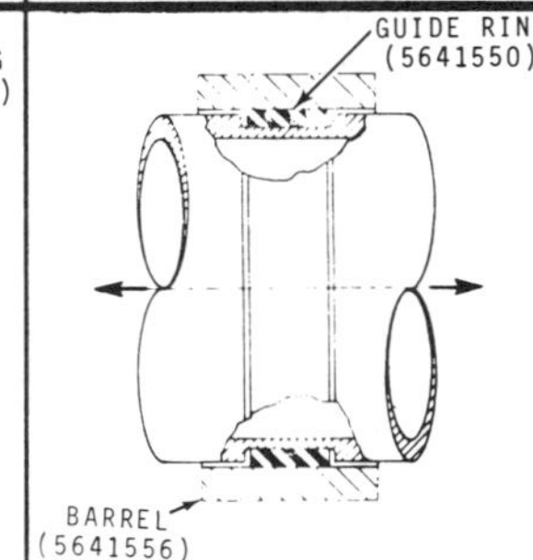	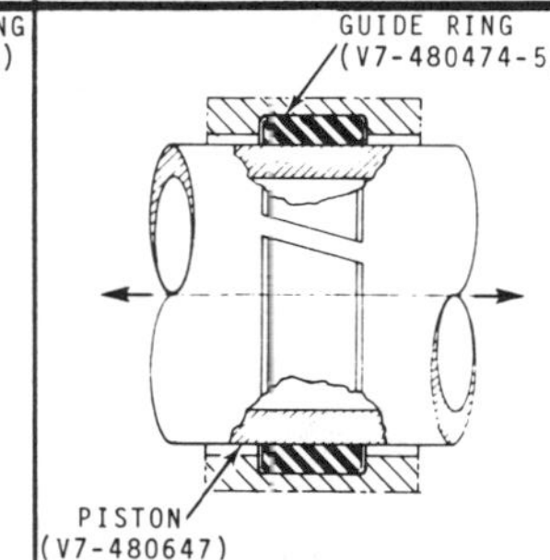	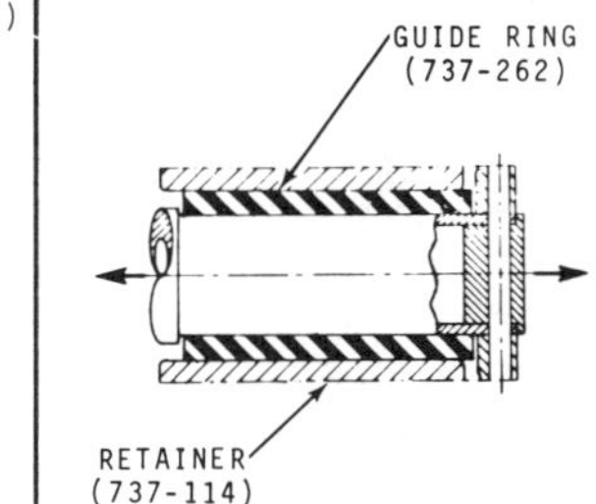
		GUIDE TYPE	SPLIT - RING	SOLID - RING	SPLIT - RING	SOLID - RING
		BASIC GUIDE BEARING DIA., IN.	.576	2.500	3.000	.187
		GUIDE WIDTH, IN.	.185	.220	.295	.385
		L/D	.322	.088	.098	1.765
		ASSOCIATED GUIDE MEMBERS	ONE OF THREE GUIDES	ONE OF THREE GUIDES	ONE OF THREE GUIDES	NONE
		DESIGN OPERATING MEDIA	LOX or LH_2	GHe	GHe	GHe or GH_2
		DESIGN OPERATING TEMP. RANGE, °F	-423 to +160	-423 to +165	-423 to +160	-423 to +160
		DESIGN OPERATING LIFE, CYCLES	750	2000	750	600
		SPECIMEN OPERATION, CYCLES	935	3318	591	1257
INTERFACE MEMBERS DESCRIPTION	GUIDE	PART NUMBER	V7-480563	5641550	V7-480474-5	737-262
		PART NAME	BEARING, PISTON SHAFT - RECIRC. VALVE	GUIDE RING	RING - FILL VALVE	SLEEVE - GUIDE ASSY
		MATERIAL	MB0130-052 TEFLON	FLUOROCOMP #105 (REIN. TEFLON)	MB0130-052 TEFLON	AMS3651 TEFLON
		SURFACE FINISH	125 r m s	125 r m s	125 r m s	125 r m s
		THICKNESS	.031	.035	.082	.032
	MATING MEMBER	PART NUMBER	V7-480642-11	5641556	V7-480647	737-114
		PART NAME	HOUSING, RECIRC. VALVE	BARREL - ACTUATOR	PISTON - ACT., FILL VALVE	RETAINER - PRIMARY PILOT SEAT
		MATERIAL	MB0170-028 TENS. 50 - T6 AL. ALY.	QQ-A-200/8 6061 - T651 AL. ALY.	MB0160-020 304L CRES	AMS5643 17-4PH CRES
		SURFACE FINISH	32 r m s	32 r m s	4 r m s	8 r m s
		SURFACE TREATMENT	MA0109-009 ANODIZE	HARD ANODIZE	THIN DENSE CHROME PLATE	PASSIVATE
		OTHER SURFACE TREATMENTS	NONE	NONE	PASSIVATE	NONE
USING COMPONENT DESCRIPTION		PROCUREMENT DOCUMENT(S) OR P/N(S)	V7-480328-91	20M32014	V7-480450-121	ME284-0030-0039
		COMPONENT NAME	2-INCH NORMALLY OPEN RECIRCULATION VALVE	2-INCH NORMALLY CLOSED SUBMERGED LH_2 SHUTOFF VALVE	8-INCH PROPELLANT FILL & DRAIN VALVE	7-INCH LH_2 TANK VENT & RELIEF VALVE
		MANUFACTURER	ROCKWELL/LAD	PARKER-HANNIFIN	ROCKWELL/LAD	AMETEK/CALMEC
		MANUFACTURER'S PART NUMBER	V7-480328-91	5640027	V7-480450-121	737-555
		USING VEHICLE(S) AND STAGE(S)	SATURN S-II	SATURN S-IC (QUALIFIED BACK-UP)	SATURN S-II	SATURN S-II
		USING SYSTEM(S)	LOX & LH_2 RECIRCULATION	DNA	LOX & LH_2 TANK FILL & DRAIN	LH_2 TANK VENT

Interface Category Number 8
(Sheet 1 of 5)

RECIPROCATING GUIDES (METAL–TO–METAL)
– Radial Bearing

		DESIGN CONCEPT NUMBER	8-1	8-2	8-3	8-4
COMPONENT DATA BOOK		CONTROL NUMBER(S)	S – IVB / 209 – 2	S – IVB / 209 – 2	S – II / 90 – 1	S – II / 96 – 5
		INTERFACE DASH NUMBER(S)	– 2	– 13	– 9	– 11
		WEAR ANALYSIS RATING (AVG.)	**1.3**	**5.6**	**6.2**	**6.3**
		PHOTOGRAPHS, PAGES	B – 1, – 2, – 4, – 7 – 8, – 9, – 10, & – 11	B – 6 & – 17	B – 4 – 16 & – 17	B – 4, – 12, & – 13
INTERFACE DESCRIPTION		INTERFACE PICTORIAL DESCRIPTION	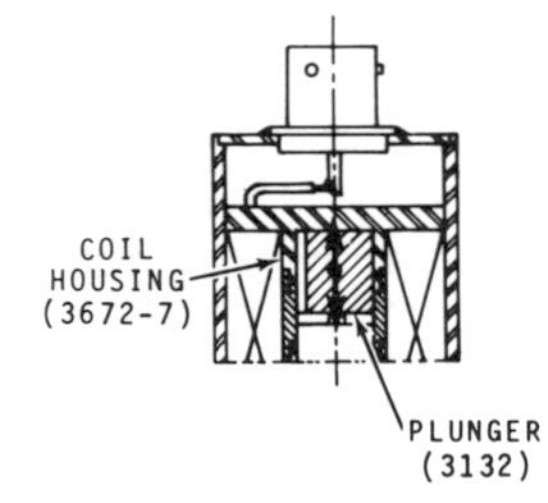	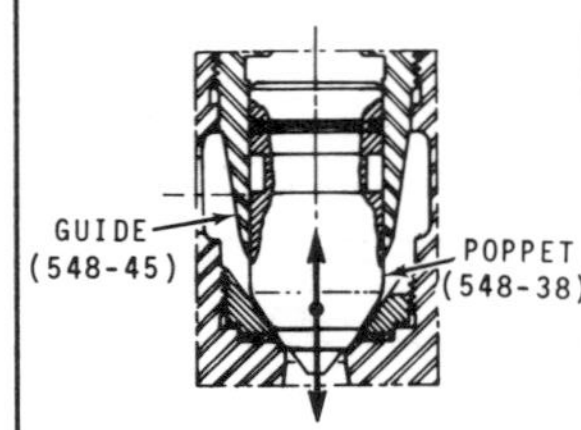	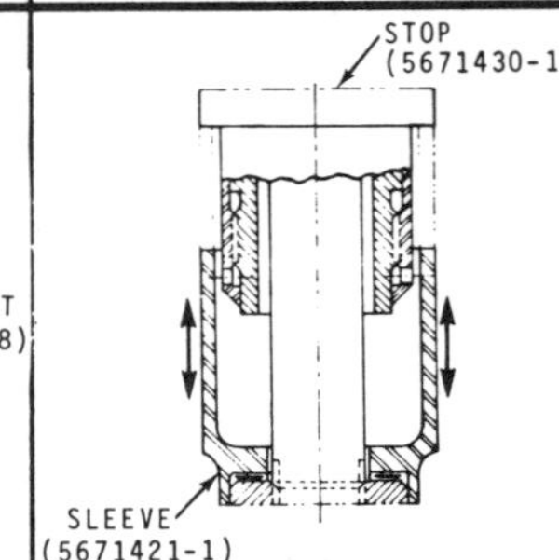	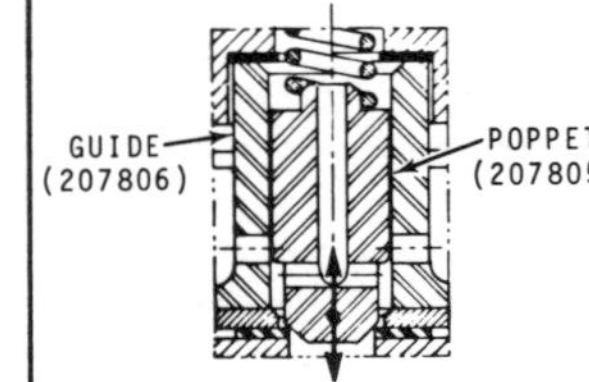
		BASIC GUIDE BEARING DIA., IN.	.4375	.700	1.070	.155
		GUIDE WIDTH, IN.	.580	.95	.240 – .920	.204
		L / D	1.330	1.35	.224 – .860	1.320
		DIAMETRAL CLEARANCE, IN.	.0043 – .0065	.0025 – .0035	.085 – .115	.0008 – .0013
		ASSOCIATED GUIDE MEMBERS	NONE	NONE	NONE	NONE
		DESIGN OPERATING MEDIA	GHe	GHe	LOX or LH_2	GHe
		DESIGN OPERATING TEMP. RANGE, °F	– 423 to + 160	– 423 to + 160	– 423 to + 160	– 180 to + 160
		DESIGN OPERATING LIFE, CYCLES	5000	5000	1000	500
		SPECIMEN OPERATION, CYCLES	10,121	10,121	1178	565
INTERFACE MEMBERS DESCRIPTION	GUIDE MEMBER NO. 1	PART NUMBER	3672 – 7	548 – 45	5671430 – 1	207806
		PART NAME	COIL HOUSING	GUIDE, VALVE POPPET	STOP, KICK – OFF	GUIDE – POPPET, METERING, REG.
		MATERIAL	1213C L. C. STEEL & 18 – 8 CRES	AMS 5643 17 – 4PH CRES	AMS 5737 A – 286 CRES	AMS 4631 ALUMINUM BRONZE
		HEAT TREATMENT	AS REC'D. MAT'L. COND.	COND. H – 925	R_c 26 MIN.	AS REC'D. MAT'L. COND.
		SURFACE FINISH	125 r m s	8 r m s	32 r m s	16 r m s
		SURFACE TREATMENT	NICKEL PLATE (1213C)	PASSIVATE	CHROME PLATE	DRY FILM LUBRICANT
		OTHER SURFACE TREATMENTS	NONE	NONE	DRY FILM LUBRICANT	NONE
	GUIDE MEMBER NO. 2	PART NUMBER	3132	548 – 38	5671421 – 1	207805
		PART NAME	PLUNGER, SOLENOID VALVE	POPPET, SOLENOID VALVE	SLEEVE, VOLUTE SPRING	POPPET – METERING, REGULATOR
		MATERIAL	1213C LOW CARBON STEEL	AMS 5643 17 – 4PH CRES	AMS 5737 A – 286 CRES	AMS 5737 A – 286 CRES
		HEAT TREATMENT	AS REC'D. MAT'L. COND.	COND. H – 925	R_C24 MIN.	COND. H. T.
		SURFACE FINISH	125 r m s	6 r m s	125 r m s	8 r m s
		SURFACE TREATMENT	NICKEL PLATE	HARD CHROME PLATE	PASSIVATE	DRY FILM LUBRICANT
		OTHER SURFACE TREATMENTS	NONE	PASSIVATE	NONE	PASSIVATE
USING COMPONENT DESCRIPTION		PROCUREMENT DOCUMENT(S) OR P/N(S)	1B43660 – 515	1B43660 – 515	ME284 – 0358 – 0001	ME284 – 0158 – 0004
		COMPONENT NAME	1 / 2 – INCH NORMALLY CLOSED HELIUM SOLENOID VALVE	1 / 2 – INCH NORMALLY CLOSED HELIUM SOLENOID VALVE	8 – INCH PROPELLANT SHUTOFF VALVE (PREVALVE)	3 / 8 – INCH HELIUM INJECTION REGULATOR – RELIEF VALVE
		MANUFACTURER	AMETEK / CALMEC	AMETEK / CALMEC	PARKER – HANNIFIN	ROYAL INDUSTRIES
		MANUFACTURER'S PART NUMBER	548 – 515	548 – 515	5670024	12078
		USING VEHICLE(S) AND STAGE(S)	SATURN S – IVB	SATURN S – IVB	SATURN S – II	SATURN S – II
		USING SYSTEM(S)	PRESSURE CONTROL	PRESSURE CONTROL	LOX & LH_2 FEED	VALVE ACTUATION, HELIUM INJECTION, & ACCUMULATOR FILL

Interface Category Number 8
(Sheet 2 of 5)

RECIPROCATING GUIDES (METAL–TO–METAL)
– Radial Bearing

	DESIGN CONCEPT NUMBER	8–5	8–6	8–7	8–8
COMPONENT DATA BOOK	CONTROL NUMBER(S)	S – IVB / 210 – 3	S – II / 104 – 2; S – IC / 1 – 2	S – IC / 1 – 2; S – II / 104 – 2	S – II / 96 – 5
	INTERFACE DASH NUMBER(S)	– 2, – 2A, & – 2B	– 1 (6.8) *; – 8 (8.2) *	– 9 (7.7) *; – 8 (7.8) *	–9
	WEAR ANALYSIS RATING (AVG.)	**7.1**	**7.5 ***	**7.7 ***	**7.8**
	PHOTOGRAPHS, PAGES	B – 2, – 3, – 4, – 5, & – 9	B – 2, & – 4; B – 2, – 8, & – 9	B – 2, & – 9; B – 2 & – 11	B – 7
INTERFACE DESCRIPTION	INTERFACE PICTORIAL DESCRIPTION	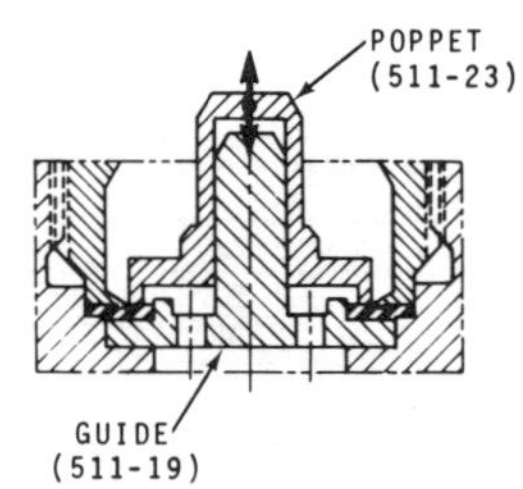	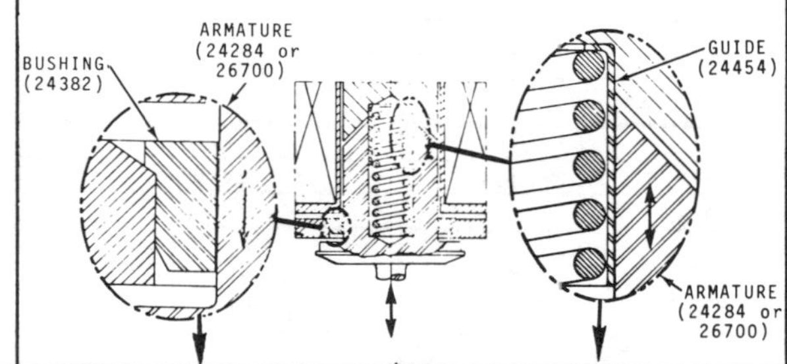		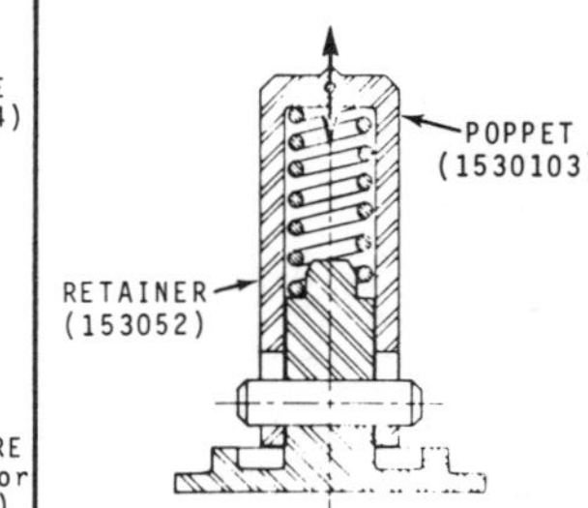
	BASIC GUIDE BEARING DIA., IN.	.092	.794	.375	.125
	GUIDE WIDTH, IN.	.153	.105	.370	.200
	L / D	1.66	.132	.986	1.600
	DIAMETRAL CLEARANCE, IN.	.002 – .004	.0025 – .0050	.001 – .008	.001 – .004
	ASSOCIATED GUIDE MEMBERS	NONE	SEE DESIGN CONCEPT 8 – 7	SEE DESIGN CONCEPT 8 – 6	SEE DESIGN CONCEPT 8 – 12
	DESIGN OPERATING MEDIA	GHe	VALVE EXTERNAL ENVIRONMENT	VALVE EXTERNAL ENVIRONMENT	GHe
	DESIGN OPERATING TEMP. RANGE, °F	–423 to + 160	– 180 to + 160; – 200 to + 165	– 200 to + 165; – 180 to + 160	– 180 to + 160
	DESIGN OPERATING LIFE, CYCLES	200	2000	2000	150
	SPECIMEN OPERATION, CYCLES	30	1096; 5248	5248; 1096	195
INTERFACE MEMBERS DESCRIPTION — GUIDE MEMBER NO. 1	PART NUMBER	511 – 19	24382	24454	153052
	PART NAME	GUIDE, CHECK VALVE POPPET	BUSHING	GUIDE, ARMATURE	RETAINER – POPPET ASSY, R.V., REG.
	MATERIAL	MIL – S – 7720 18 – 8 / 303 CRES	MIL – S – 7720 303 CRES	MIL – T – 8808 321 CRES	QQ – A – 268 2024 – T351 AL. ALY.
	HEAT TREATMENT	COND. B	COND. A	AS REC'D. TUBE	COND. – T351
	SURFACE FINISH	16 r m s	32 r m s	AS REC'D SEAMLESS TUBE	125 r m s
	SURFACE TREATMENT	PASSIVATE	PASSIVATE	PASSIVATE	DRY FILM LUBRICANT
	OTHER SURFACE TREATMENTS	NONE	NONE	NONE	ANODIZE
INTERFACE MEMBERS DESCRIPTION — GUIDE MEMBER NO. 2	PART NUMBER(S)	511 – 23	24284; 26700	26700; 24284	1530103
	PART NAME	POPPET, CHECK VALVE	ARMATURE	ARMATURE	POPPET – POPPET ASSY, R.V., REG.
	MATERIAL	MIL – S – 7720 18 – 8 / 303 CRES	AISI C12L14 COLD FIN. LEDLOY # 300	AISI C12L14 COLD FIN. LEDLOY # 300	AMS 5737 A – 286 CRES
	HEAT TREATMENT	COND. B	SPECIFIED H. T. CYCLE	SPECIFIED H. T. CYCLE	COND. H. T.
	SURFACE FINISH	16 r m s	32 r m s	32 r m s	125 r m s
	SURFACE TREATMENT	PASSIVATE	ELECTROLESS NICKEL PLATE	ELECTROLESS NICKEL PLATE	DRY FILM LUBRICANT
	OTHER SURFACE TREATMENTS	NONE	NONE; DRY FILM LUBRICANT	DRY FILM LUBRICANT; NONE	PASSIVATE
USING COMPONENT DESCRIPTION	PROCUREMENT DOCUMENT(S) OR P/N(S)	1A49988	ME284 – 0356 – 0001; 60B52101 – 15F	60B52101 – 15F ME284 – 0356 – 0001	ME284 – 0158 – 0004
	COMPONENT NAME	6 – INCH DIRECTIONAL CONTROL FUEL VENT VALVE	3 / 8 – INCH NORMALLY CLOSED SOLENOID VALVE	3 / 8 – INCH NORMALLY CLOSED SOLENOID VALVE	3 / 8 – INCH HELIUM INJECTION REGULATOR – RELIEF VALVE
	MANUFACTURER	AMETEK / CALMEC	J. C. CARTER	J. C. CARTER	ROYAL INDUSTRIES
	MANUFACTURER'S PART NUMBER	511 – 509	6969 – 2; 6969 – 15	6969 – 15; 6969 – 2	12078
	USING VEHICLE(S) AND STAGE(S)	SATURN S – IVB	SATURN S – II; SATURN S – IC	SATURN S – IC; SATURN S – II	SATURN S – II
	USING SYSTEM(S)	LH_2 TANK VENT	VALVE ACTUATION, HELIUM INJECTION, & ACCUMULATOR FILL	VALVE ACTUATION, HELIUM INJECTION, & ACCUMULATOR FILL	VALVE ACTUATION, HELIUM INJECTION, & ACCUMULATOR FILL

Interface Category Number 8 (Sheet 3 of 5)

RECIPROCATING GUIDES (METAL-TO-METAL) – Radial Bearing

		DESIGN CONCEPT NUMBER	8-9	8-10	8-11	8-12
COMPONENT DATA BOOK		CONTROL NUMBER(S)	S – IVB / 209 – 2	S – IVB / 210 – 3	S – II / 100 – 1	S – II / 96 – 5
		INTERFACE DASH NUMBER(S)	– 3	– 5 & – 5A	– 12	– 9
		WEAR ANALYSIS RATING (AVG.)	**7.9**	**8.1**	**8.3**	**8.5**
		PHOTOGRAPHS, PAGES	B – 4, & – 12	B – 3, – 4, – 5, & – 10	81, & 94	B – 7
INTERFACE DESCRIPTION		INTERFACE PICTORIAL DESCRIPTION	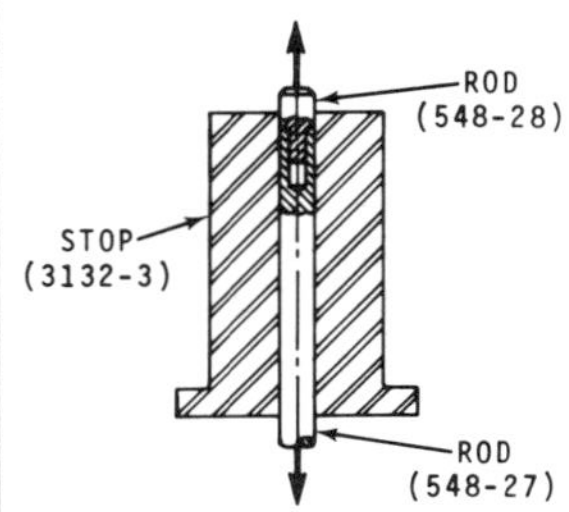	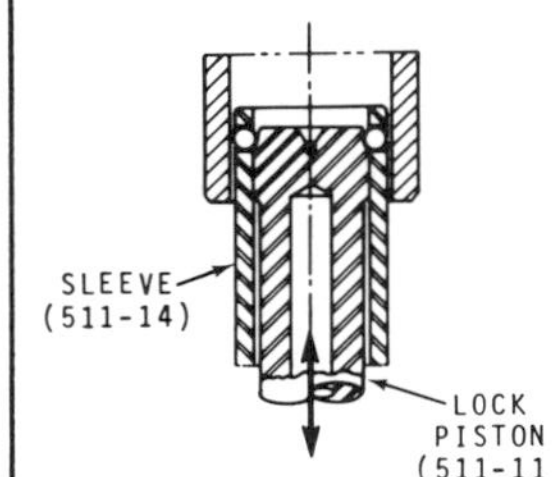	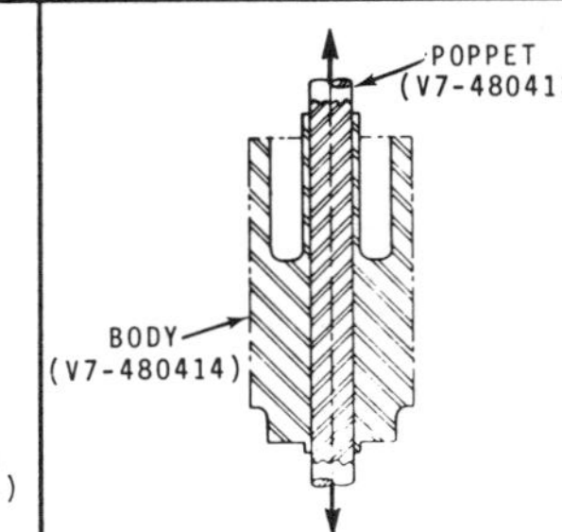	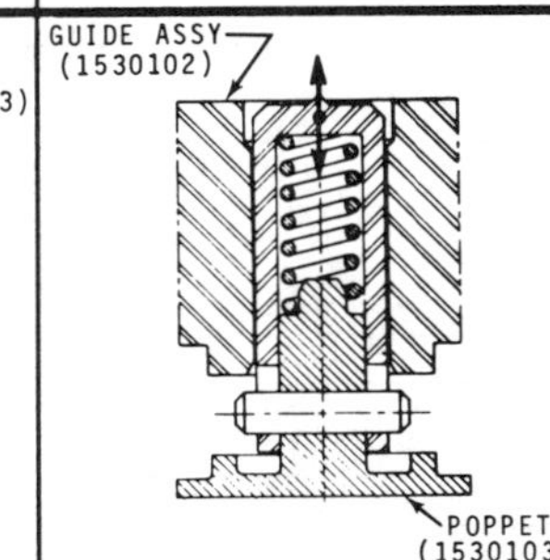
		BASIC GUIDE BEARING DIA., IN.	.090	.602	.113	.187
		GUIDE WIDTH, IN.	.80	.165	.937	.300
		L / D	8.90	.274	8.820	1.600
		DIAMETRAL CLEARANCE, IN.	.005 – .007	.003 – .006	.0020 – .0035	.0005 – .0012
		ASSOCIATED GUIDE MEMBERS	NONE	SEE DESIGN CONCEPT 8 – 15	NONE	SEE DESIGN CONCEPT 8 – 8
		DESIGN OPERATING MEDIA	GHe	GHe	LOX or LH_2	GHe
		DESIGN OPERATING TEMP. RANGE, °F	– 423 to + 160	– 423 to + 160	– 423 to + 160	– 180 to + 160
		DESIGN OPERATING LIFE, CYCLES	5000	200	750	150
		SPECIMEN OPERATION, CYCLES	10,121	30	935	195
INTERFACE MEMBERS DESCRIPTION	GUIDE MEMBER NO. 1	PART NUMBER	3132 – 3	511 – 14	V7 – 480414	1530102
		PART NAME	STOP, SOLENOID PLUNGER	SLEEVE, BALL, RETAINER	BODY – VALVE, RELIEF, ASSY OF	GUIDE ASSY – POPPET, R. V., REG.
		MATERIAL	QQ – S – 637 C – 1213 COLD FORM, STL	AMS 5643 17 – 4PH CRES	QQ – 763 303 Se CRES	QQ – S – 763 303 Se CRES
		HEAT TREATMENT	AS REC'D. MAT'L COND.	COND. H – 925	COND. A	COND. B
		SURFACE FINISH	32 r m s	32 r m s	16 r m s	16 r m s
		SURFACE TREATMENT	UNKNOWN	PASSIVATE	MICROSEAL PROCESS #100 – 1	PASSIVATE
		OTHER SURFACE TREATMENTS	NONE	NONE	NONE	NONE
	GUIDE MEMBER NO. 2	PART NUMBER(S)	548 – 27 & 548 – 28	511 – 11	V7 – 480413	1530103
		PART NAME(S)	ROD, SOLENOID PUSH & ARMATURE	PISTON, LOCK	POPPET – RELIEF VALVE	POPPET – POPPET ASSY, R. V. REG.
		MATERIAL	MIL – S – 7720 18 – 8 / 303 CRES	AMS 5643 17 – 4 PH CRES	QQ – S – 763 303 Se CRES	AMS 5737 A – 286 CRES
		HEAT TREATMENT	COND. B	COND. H – 925	COND. A	COND. H. T.
		SURFACE FINISH	32 r m s	32 r m s	16 r m s	16 r m s
		SURFACE TREATMENT	PASSIVATE	DENSE CHROME PLATE	PASSIVATE	DRY FILM LUBRICANT
		OTHER SURFACE TREATMENTS	NONE	PASSIVATE	NONE	PASSIVATE
USING COMPONENT DESCRIPTION		PROCUREMENT DOCUMENT(S) OR P/N(S)	1B43660 – 515	1A49988	V7 – 480328 – 91	ME284 – 0158 – 0004
		COMPONENT NAME	1/2 – INCH NORMALLY CLOSED HELIUM SOLENOID VALVE	6 – INCH DIRECTIONAL CONTROL FUEL VENT VALVE	2 – INCH NORMALLY OPEN RECIRCULATION VALVE	3 / 8 – INCH HELIUM INJECTION REGULATOR – RELIEF VALVE
		MANUFACTURER	AMETEK / CALMEC	AMETEK / CALMEC	ROCKWELL / LAD	ROYAL INDUSTRIES
		MANUFACTURER'S PART NUMBER	548 – 515	511 – 509	V7 – 480328 – 91	12078
		USING VEHICLE(S) AND STAGE(S)	SATURN S – IVB	SATURN S – IVB	SATURN S – II	SATURN S – II
		USING SYSTEM(S)	PRESSURE CONTROL	LH_2 TANK VENT	LOX & LH_2 RECIRCULATION	VALVE ACTUATION, HELIUM INJECTION, & ACCUMULATOR FILL

Interface Category Number 8
(Sheet 4 of 5)

RECIPROCATING GUIDES (METAL-TO-METAL)
- Radial Bearing

	DESIGN CONCEPT NUMBER	8-13	8-14	8-15	8-16
COMPONENT DATA BOOK	CONTROL NUMBER(S)	S - II / 92 - 3	S - IVB / 210 - 3	S - IVB / 210 - 3	S - IC / 207 - 5
	INTERFACE DASH NUMBER(S)	- 13	- 8 & - 8A	- 4 & - 4A	- 22
	WEAR ANALYSIS RATING (AVG.)	**8.5**	**8.8**	**9.3**	**9.3**
	PHOTOGRAPHS, PAGES	87, 102, 103	B - 3, - 4, - 5, - 6, - 11, & - 12	B - 1, - 3, - 4, - 5, & - 10	B - 10, & - 23
INTERFACE DESCRIPTION	INTERFACE PICTORIAL DESCRIPTION	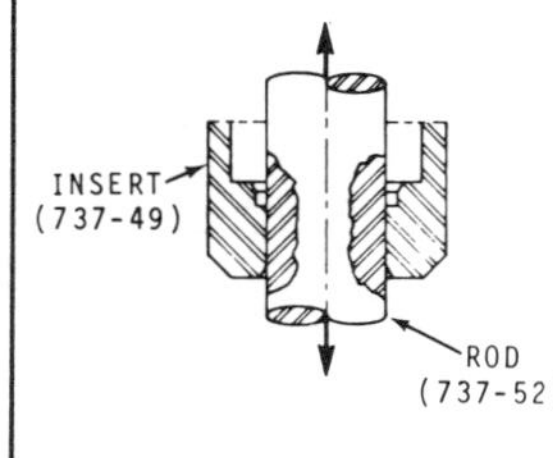	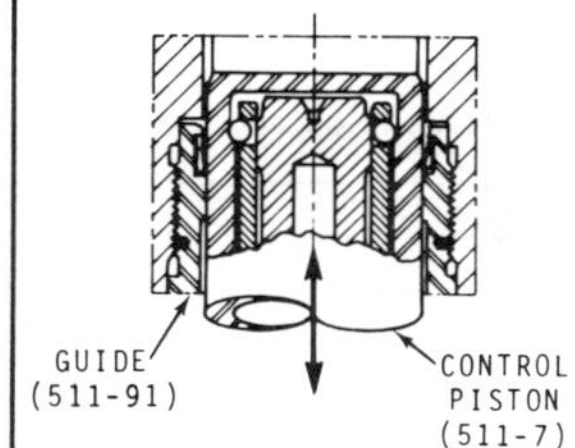	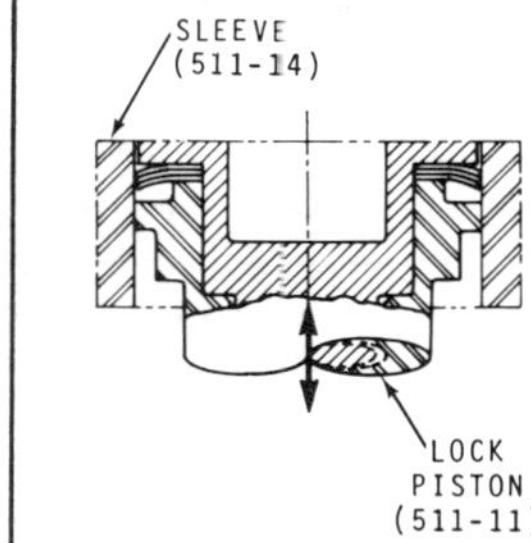	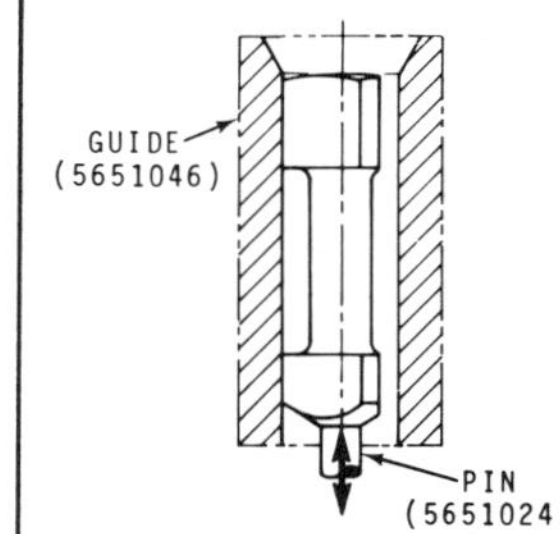
	BASIC GUIDE BEARING DIA., IN.	.3125	1.125	.811	.1490
	GUIDE WIDTH, IN.	.150	.246	.068	.450
	L / D	.480	.219	.084	3.00
	DIAMETRAL CLEARANCE, IN.	.0010 - .0025	.0010 - .0020	.002 - .004	.0010 - .0025
	ASSOCIATED GUIDE MEMBERS	SEE DESIGN CONCEPT 2 - 8	TWO GUIDES ON 4.40 IN. CENTERS	SEE DESIGN CONCEPT 8 - 10	NONE
	DESIGN OPERATING MEDIA	GHe	GHe	GHe	GHe, GN_2, or RP - 1
	DESIGN OPERATING TEMP. RANGE, °F	- 423 to + 160	- 423 to + 160	- 423 to + 160	- 400 to + 160
	DESIGN OPERATING LIFE, CYCLES	600	200	200	500
	SPECIMEN OPERATION, CYCLES	1257	30	30	500
INTERFACE MEMBERS DESCRIPTION — GUIDE MEMBER NO. 1	PART NUMBER	737 - 49	511 - 91	511 - 14	5651046
	PART NAME	INSERT - OVERRIDE PISTON	GUIDE, CONTROL PISTON	SLEEVE, BALL RETAINER	GUIDE, METERING PIN
	MATERIAL	MIL - S - 7720 18 - 8 / 303 CRES	AMS 5643 17 - 4 PH CRES	AMS 5643 17 - 4 PH CRES	QQ - S - 763 303 Se CRES
	HEAT TREATMENT	AS REC'D, MAT'L COND.	COND. H - 925	COND. H - 925	COND. B
	SURFACE FINISH	16 r m s	8 r m s	32 r m s	32 r m s
	SURFACE TREATMENT	PASSIVATE	"LUBE - LOCK #4396"	PASSIVATE	PASSIVATE
	OTHER SURFACE TREATMENTS	NONE	PASSIVATE	NONE	NONE
GUIDE MEMBER NO. 2	PART NUMBER	737 - 52	511 - 7	511 - 11	5651024
	PART NAME	ROD - OVERRIDE PISTON	PISTON, CONTROL	PISTON, LOCK	PIN, METERING
	MATERIAL	MIL - S - 7720 18 - 8 / 303 CRES	AMS 6543 17 - 4 PH CRES	AMS 5643 17 - 4PH CRES	AMS 5737 A - 286 CRES
	HEAT TREATMENT	COND. B	COND. H - 925	COND. H - 925	AS REC'D. MAT'L COND.
	SURFACE FINISH	16 r m s	8 r m s	32 r m s	32 r m s
	SURFACE TREATMENT	HARD CHROME PLATE	HARD CHROME PLATE	DENSE CHROME PLATE	PASSIVATE
	OTHER SURFACE TREATMENTS	PASSIVATE	PASSIVATE	PASSIVATE	NONE
USING COMPONENT DESCRIPTION	PROCUREMENT DOCUMENT(S) OR P/N(S)	ME284 - 0030 - 0039	1A49988	1A49988	20M32015
	COMPONENT NAME	7 - INCH LH_2 TANK VENT & RELIEF VALVE	6 - INCH DIRECTIONAL CONTROL FUEL VENT VALVE	6 - INCH DIRECTIONAL CONTROL FUEL VENT VALVE	1 - 1 / 2 INCH HELIUM PRESSURE REGULATOR
	MANUFACTURER	AMETEK / CALMEC	AMETEK / CALMEC	AMETEK / CALMEC	PARKER - HANNIFIN
	MANUFACTURER'S PART NUMBER	737 - 555	511 - 509	511 - 509	5640061
	USING VEHICLE(S) AND STAGE(S)	SATURN S - II	SATURN S - IVB	SATURN S - IVB	SATURN S - IC (QUALIFIED BACK - UP)
	USING SYSTEM(S)	LH_2 TANK VENT	LH_2 TANK VENT	LH_2 TANK VENT	DNA

Interface Category Number 8 (Sheet 5 of 5)

RECIPROCATING GUIDES (METAL-TO-METAL) - Radial Bearing

DESIGN CONCEPT NUMBER		8-17			
COMPONENT DATA BOOK	CONTROL NUMBER(S)	S-IVB/209-2			
	INTERFACE DASH NUMBER(S)	-9			
	WEAR ANALYSIS RATING (AVG.)	**9.6**			
	PHOTOGRAPHS, PAGES	B-5, -6, & -15			
INTERFACE DESCRIPTION	INTERFACE PICTORIAL DESCRIPTION	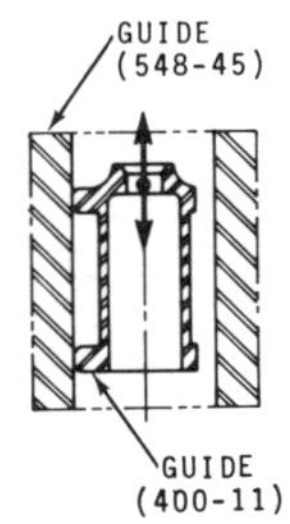			
	BASIC GUIDE BEARING DIA., IN.	.187			
	GUIDE WIDTH, IN.	.240			
	L/D	1.280			
	DIAMETRAL CLEARANCE, IN.	.001 - .003			
	ASSOCIATED GUIDE MEMBERS	TWO .032 WIDE GUIDES			
	DESIGN OPERATING MEDIA	GHe			
	DESIGN OPERATING TEMP. RANGE, °F	-423 to +160			
	DESIGN OPERATING LIFE, CYCLES	5000			
	SPECIMEN OPERATION, CYCLES	10,121			
INTERFACE MEMBERS DESCRIPTION — GUIDE MEMBER NO. 1	PART NUMBER	548-45			
	PART NAME	GUIDE, VALVE POPPET			
	MATERIAL	AMS 5643 17-4 PH CRES			
	HEAT TREATMENT	COND. H-925			
	SURFACE FINISH	32 rms			
	SURFACE TREATMENT	PASSIVATE			
	OTHER SURFACE TREATMENTS	NONE			
GUIDE MEMBER NO. 2	PART NUMBER	400-11			
	PART NAME	GUIDE, PILOT POPPET SOL. VALVE			
	MATERIAL	MIL-S-7720 18-8/303 CRES			
	HEAT TREATMENT	COND. B			
	SURFACE FINISH	16 rms			
	SURFACE TREATMENT	PASSIVATE			
	OTHER SURFACE TREATMENTS	NONE			
USING COMPONENT DESCRIPTION	PROCUREMENT DOCUMENT(S) OR P/N(S)	1B43660-515			
	COMPONENT NAME	1/2-INCH NORMALLY CLOSED HELIUM SOLENOID VALVE			
	MANUFACTURER	AMETEK/CALMEC			
	MANUFACTURER'S PART NUMBER	548-515			
	USING VEHICLE(S) AND STAGE(S)	SATURN S-IVB			
	USING SYSTEM(S)	PRESSURE CONTROL			

Interface Category Number 9
(Sheet 1 of 2)

ROTATING SHAFT JOURNAL BEARINGS
– Radial Bearing

		DESIGN CONCEPT NUMBER	9-1	9-2	9-3	9-4
COMPONENT DATA BOOK		CONTROL NUMBER(S)	S – II / 90 – 1	S – II / 90 – 1	S – I / 89 – 1	S – II / 89 – 1
		INTERFACE DASH NUMBER(S)	– 5	– 3	– 19	– 2 & – 6
		WEAR ANALYSIS RATING (AVG.)	**7.9**	**8.5**	**8.7**	**8.8**
		PHOTOGRAPHS, PAGES	B – 2, & B – 21	B – 2, – 11, & – 20	B – 2, – 3, & – 20	B – 2, – 3, – 8, & – 9
INTERFACE DESCRIPTION		INTERFACE PICTORIAL DESCRIPTION	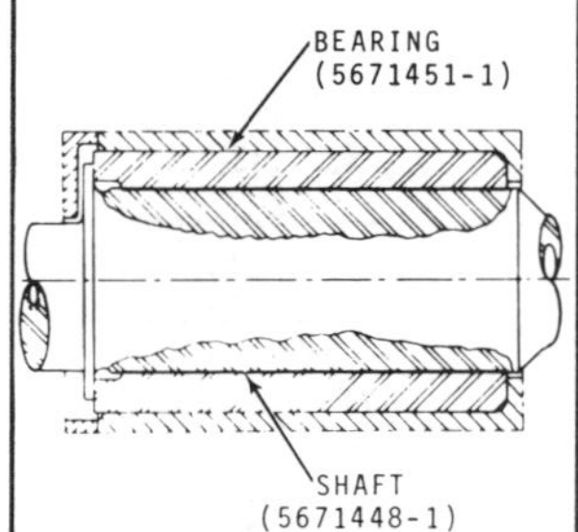	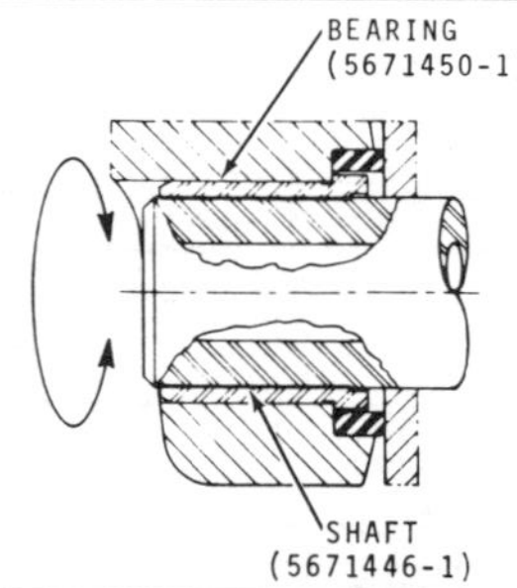	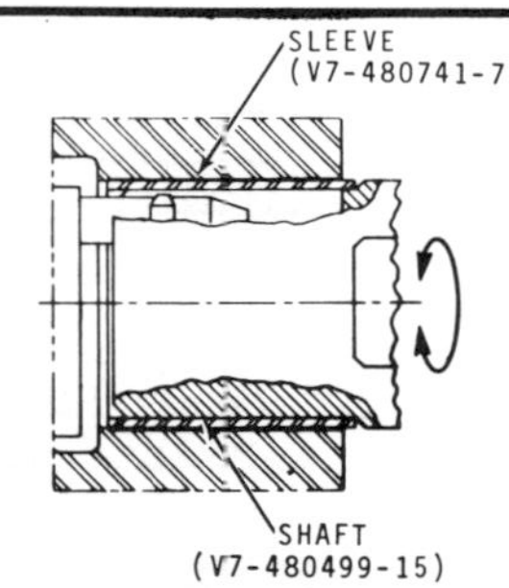	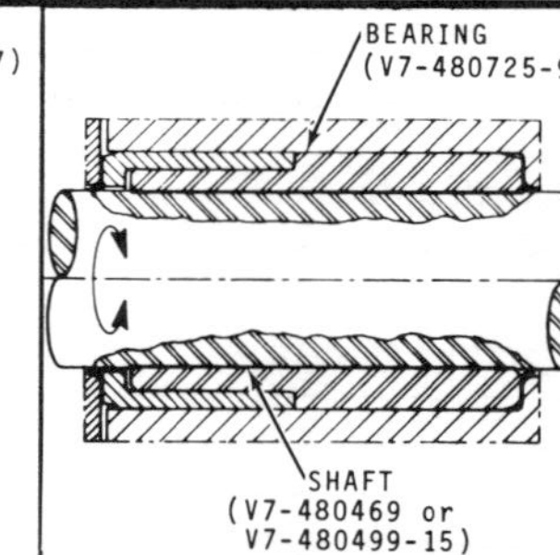
		BEARING TYPE	SOLID METAL	SOLID METAL	SOLID METAL	SOLID METAL
		BASIC BEARING DIA. / LENGTH, IN.	1.000 / 2.100	1.000 / .960	.650 / .640	1.000 / 2.175
		L / D	2.100	.960	.985	2.175
		DIAMETRAL CLEARANCE, IN.	.0025 – .0035	.0001 – .0011	.0010 – .0018	.0013 – .0020
		ASSOCIATED BEARING MEMBERS	TWO BALL BEARING UNITS	NONE	SEE DESIGN CONCEPT 9 – 4	SEE DESIGN CONCEPT 9 – 3
		DESIGN OPERATING MEDIA	LOX or LH_2	LOX or LH_2	AIR (GOX or GH_2)	AIR (GOX or GH_2)
		DESIGN OPERATING TEMP. RANGE, °F	– 423 to + 160	– 423 to + 160	– 423 to + 160	– 423 to + 160
		DESIGN OPERATING LIFE, CYCLES	1000	1000	750	750
		SPECIMEN OPERATION, CYCLES	1178	1178	591	591
INTERFACE MEMBERS DESCRIPTION	BEARING	PART NUMBER	5671451 – 1	5671450 – 1	V7 – 480741 – 7 (– 11) ASSY)	V7 – 480725 – 9
		PART NAME	BEARING, DRIVE SHAFT	BEARING, IDLER SHAFT	SLEEVE (ADAPTOR)	BEARING – DRIVE SHAFT
		MATERIAL	QQ – C – 530 BERYLLIUM COPPER	QQ – C – 530 BERYLLIUM COPPER	QQ – C – 530 BERYLLIUM COPPPER	QQ – C – 530 BERYLLIUM COPPER
		HEAT TREATMENT	115,000 to 135,000 PSI	115,000 to 135,000 PSI	180,000 PSI MIN.	115,000 to 135,000 PSI
		SURFACE FINISH	125 r m s	125 r m s	3 r m s	16 r m s
		SURFACE TREATMENT	SILVER PLATE	SILVER PLATE	SILVER PLATE	SILVER PLATE
		OTHER SURFACE TREATMENTS	NONE	NONE	NONE	NONE
	MATING MEMBER	PART NUMBER(S)	5671448 – 1	5671446 – 1	V7 – 480499 – 15 (– 31 ASSY)	V7 – 480469; V7 – 480499 – 31
		PART NAME(S)	SHAFT – DRIVE	SHAFT, IDLER	SLEEVE (DRIVE SHAFT)	IDLER SHAFT; DRIVE SHAFT
		MATERIAL	AMS5737 A – 286 CRES	AMS5737 A – 286 CRES	MB0170 – 041 INCONEL 718	MB0170 – 041 INCONEL 718
		HEAT TREATMENT	Rc 30 – 39	Rc 30 MIN.	175,000 PSI MIN.	175,000 PSI MIN.
		SURFACE FINISH	8 r m s	8 r m s	125 r m s	8 r m s
		SURFACE TREATMENT	CHROME PLATE	CHROME PLATE	THIN DENSE CHROME PLATE	THIN DENSE CHROME PLATE
		OTHER SURFACE TREATMENTS	NONE	NONE	PASSIVATE	NONE
USING COMPONENT DESCRIPTION		PROCUREMENT DOCUMENT(S) OR P/N(S)	ME284 – 0358 – 0001	ME284 – 0358 – 0001	V7 – 480450 – 121	V7 – 480450 – 121
		COMPONENT NAME	8 – INCH PROPELLANT SHUTOFF VALVE (PREVALVE)	8 – INCH PROPELLANT SHUTOFF VALVE (PREVALVE)	8 – INCH PROPELLANT FILL & DRAIN VALVE	8 – INCH PROPELLANT FILL & DRAIN VALVE
		MANUFACTURER	PARKER – HANNIFIN	PARKER – HANNIFIN	ROCKWELL / LAD	ROCKWELL / LAD
		MANUFACTURER'S PART NUMBER	5670024	5670024	V7 – 480450 – 121	V7 – 480450 – 121
		USING VEHICLE(S) AND STAGE(S)	SATURN S – II	SATURN S – II	SATURN S – II	SATURN S – II
		USING SYSTEM(S)	LOX & LH_2 FEED	LOX & LH_2 FEED	LOX & LH_2 TANK FILL & DRAIN	LOX & LH_2 TANK FILL & DRAIN

Interface Category Number 9
(Sheet 2 of 2)

ROTATING SHAFT JOURNAL BEARINGS
- Radial Bearing

		DESIGN CONCEPT NUMBER	9-5	9-6		
COMPONENT DATA BOOK		CONTROL NUMBER(S)	S – IVB / 210 – 3	S – II / 100 – 1		
COMPONENT DATA BOOK		INTERFACE DASH NUMBER(S)	– 21	– 3 & – 1A – 4		
COMPONENT DATA BOOK		**WEAR ANALYSIS RATING (AVG.)**	**9.0**	**9.2**		
COMPONENT DATA BOOK		PHOTOGRAPHS, PAGES	B – 8, & – 19	81, & 87		
INTERFACE DESCRIPTION		INTERFACE PICTORIAL DESCRIPTION	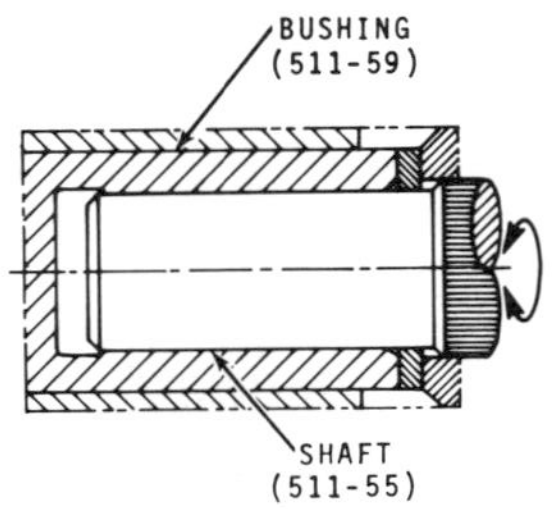	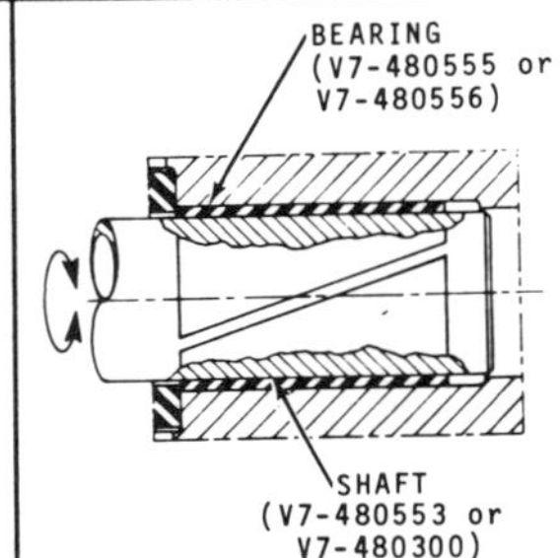		
INTERFACE DESCRIPTION		BEARING TYPE	SOLID METAL	SPLIT – RING PLASTIC		
INTERFACE DESCRIPTION		BASIC BEARING DIA. / LENGTH, IN.	.408 / .700	.420 / .705 – 1.295		
INTERFACE DESCRIPTION		L / D	1.72	1.680 – 3.080		
INTERFACE DESCRIPTION		DIAMETRAL CLEARANCE, IN.	.0020 – .0030	.002 – .006		
INTERFACE DESCRIPTION		ASSOCIATED BEARING MEMBERS	BALL BEARING UNITS	NONE		
INTERFACE DESCRIPTION		DESIGN OPERATING MEDIA	GH2 or LH2	LOX or LH2		
INTERFACE DESCRIPTION		DESIGN OPERATING TEMP. RANGE, °F	– 423 to + 160	– 423 to + 160		
INTERFACE DESCRIPTION		DESIGN OPERATING LIFE, CYCLES	200	750		
INTERFACE DESCRIPTION		SPECIMEN OPERATION, CYCLES	30	935		
INTERFACE MEMBERS DESCRIPTION	BEARING	PART NUMBER(S)	511 – 59	V7 – 480555; V7 – 480556		
INTERFACE MEMBERS DESCRIPTION	BEARING	PART NAME(S)	BUSHING, MAIN SHAFT END	BRG., IDLER; BRG., DRIVE		
INTERFACE MEMBERS DESCRIPTION	BEARING	MATERIAL	AMS5643 17 – 4PH CRES	MB0130 – 052 TEFLON		
INTERFACE MEMBERS DESCRIPTION	BEARING	HEAT TREATMENT	COND. H – 925	NONE		
INTERFACE MEMBERS DESCRIPTION	BEARING	SURFACE FINISH	8 r m s	125 r m s		
INTERFACE MEMBERS DESCRIPTION	BEARING	SURFACE TREATMENT	LUBE LOCK # 4396	NONE		
INTERFACE MEMBERS DESCRIPTION	BEARING	OTHER SURFACE TREATMENTS	PASSIVATE	NONE		
INTERFACE MEMBERS DESCRIPTION	MATING MEMBER	PART NUMBER(S)	511 – 55	V7 – 480553; V7 – 480300		
INTERFACE MEMBERS DESCRIPTION	MATING MEMBER	PART NAME(S)	SHAFT – MAIN	IDLER SHAFT; DRIVE SHAFT		
INTERFACE MEMBERS DESCRIPTION	MATING MEMBER	MATERIAL	AMS5667 INCONEL X – 750	A – 286 CRES; BeCu		
INTERFACE MEMBERS DESCRIPTION	MATING MEMBER	HEAT TREATMENT	AGED	NONE; 180,000 PSI MIN.		
INTERFACE MEMBERS DESCRIPTION	MATING MEMBER	SURFACE FINISH	8 r m s	32 r m s		
INTERFACE MEMBERS DESCRIPTION	MATING MEMBER	SURFACE TREATMENT	CHROME PLATE	PASSIVATE; THIN DENSE CHROME		
INTERFACE MEMBERS DESCRIPTION	MATING MEMBER	OTHER SURFACE TREATMENTS	PASSIVATE	NONE		
USING COMPONENT DESCRIPTION		PROCUREMENT DOCUMENT(S) OR P/N(S)	1A49988	V7 – 480328 – 91		
USING COMPONENT DESCRIPTION		COMPONENT NAME	6 – INCH DIRECTIONAL CONTROL FUEL VENT VALVE	2 – INCH NORMALLY OPEN RECIRCULATION VALVE		
USING COMPONENT DESCRIPTION		MANUFACTURER	AMETEK / CALMEC	ROCKWELL / LAD		
USING COMPONENT DESCRIPTION		MANUFACTURER'S PART NUMBER	511 – 509	V7 – 480328 – 91		
USING COMPONENT DESCRIPTION		USING VEHICLE(S) AND STAGE(S)	SATURN S – IVB	SATURN S – II		
USING COMPONENT DESCRIPTION		USING SYSTEM(S)	LH_2 TANK VENT	LOX & LH_2 RECIRCULATION		

Interface Category Number 10
(Sheet 1 of 3)

ROTATING SHAFT BALL AND ROLLER BEARINGS AND BALL SCREWS – Radial Bearing

		10-1	10-2	10-3	10-4
	DESIGN CONCEPT NUMBER	10-1	10-2	10-3	10-4
COMPONENT DATA BOOK	CONTROL NUMBER(S)	S – II / 90 – 1	S – II / 90 – 1	S – II / 98 – 5	S – IC / 207 – 5
COMPONENT DATA BOOK	INTERFACE DASH NUMBER(S)	– 6A & – 6B	– 6 & – 6C	– 14	– 5
COMPONENT DATA BOOK	**WEAR ANALYSIS RATING (AVG.)**	**4.9**	**5.2**	**6.3**	**7.9**
COMPONENT DATA BOOK	PHOTOGRAPHS, PAGES	B – 5, – 6, – 12, – 13, – 14, – 15, & – 20	B – 5, – 6, – 12, – 13, – 14, – 15, & – 20	B – 3, – 4, & – 18	B – 4, – 5, – 6, & – 16
INTERFACE DESCRIPTION	INTERFACE PICTORIAL DESCRIPTION	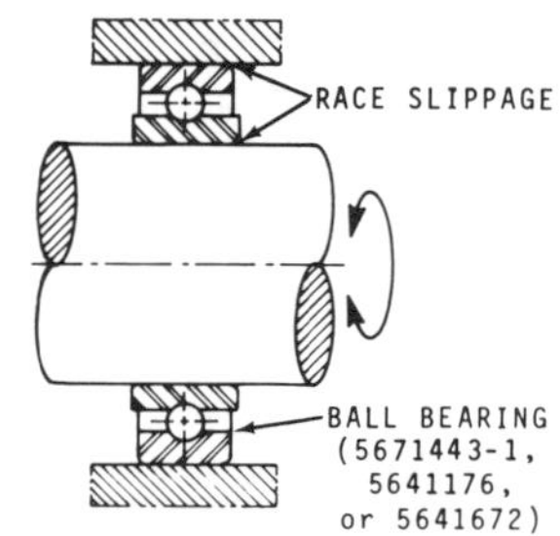			
INTERFACE DESCRIPTION	BEARING TYPE	BALL	BALL	BALL	BALL
INTERFACE DESCRIPTION	OUTER RACE O. D., IN.	2.5620 ± .0005	2.5620 ± .0005	1.6250 + .0000, – .0005	3.0000 + .0000, – .0005
INTERFACE DESCRIPTION	INNER RACE I. D., IN.	1.8125 + .0005, – .0000	1.8125 + .0005, – .0000	.4820 + .0000, – .0005	2.0000 + .0000, – .0005
INTERFACE DESCRIPTION	OUTER RACE DIA. CLEARANCE, IN.	.0005 – .0025	.0005 – .0025	.0005 – .0013	.0005 – .0020
INTERFACE DESCRIPTION	INNER RACE DIA. CLEARANCE, IN.	.0010 – .0025	.0010 – .0025	.0005 – .0013	.0000 – .0015
INTERFACE DESCRIPTION	DESIGN OPERATING MEDIA	LOX or LH_2	LOX or LH_2	GOX	GHe, GN_2 or AIR
INTERFACE DESCRIPTION	DESIGN OPERATING TEMP. RANGE, °F	– 423 to + 160	– 423 to + 160	– 180 to + 860	– 400 to + 165
INTERFACE DESCRIPTION	DESIGN OPERATING LIFE, CYCLES	1000	1000	2000	500
INTERFACE DESCRIPTION	SPECIMEN OPERATION, CYCLES	1178	1178	147	500
INTERFACE MEMBERS DESCRIPTION – BEARING	PART NUMBER	5671443 – 1	5671443 – 1	5641176	5641672
INTERFACE MEMBERS DESCRIPTION – BEARING	PART NAME	BEARING – CONNECTING ROD	BEARING – CONNECTING ROD	BEARING – RADIAL, LOWER	BEARING – RADIAL, DEEP GROOVE
INTERFACE MEMBERS DESCRIPTION – BEARING	MANUFACTURER P/N	RAYTEE P / N 46650B	FAFNIR P / N AK29B5	INDUSTRIAL TECHTRONICS P / N 7709	INDUSTRIAL TECHTRONICS P / N 8652
INTERFACE MEMBERS DESCRIPTION – BEARING	RACES MATERIAL	QQ – S – 763 440C CRES	QQ – S – 763 440C CRES	STELLITE 19	QQ – S – 763 440C CRES
INTERFACE MEMBERS DESCRIPTION – BEARING	HEAT TREATMENT	R_c 58 MIN.	R_c 58 MIN.	AS REC'D. MAT'L.	R_c 55 MIN.
INTERFACE MEMBERS DESCRIPTION – BEARING	SURFACE FINISH	16 r m s	16 r m s	125 r m s	125 r m s
INTERFACE MEMBERS DESCRIPTION – BEARING	SURFACE TREATMENT	PASSIVATE	PASSIVATE	NONE	CHROME PLATE
INTERFACE MEMBERS DESCRIPTION – MATING MEMBERS	PART NUMBER(S)	5671434 – 1; 5671435 – 1	5671436 – 1; 5671435 – 1 ; 5671438 – 1	5641006; 5641525	5651002; 5651013
INTERFACE MEMBERS DESCRIPTION – MATING MEMBERS	PART NAME(S)	ROD; CRANK	INSERT, CRANK; COLLAR	BODY; SHAFT	BALL SCREW ASSY; HOUSING
INTERFACE MEMBERS DESCRIPTION – MATING MEMBERS	MATERIAL	INCONEL X750; 15 – 5 PH CRES	INVAR 36; 15 – 5 PH CRES	17 – 4PH CRES; A – 286 CRES	PH13 – 8Mo; CRES
INTERFACE MEMBERS DESCRIPTION – MATING MEMBERS	HEAT TREATMENT	R_c 30 MIN.; RHc 24/30	AS REC'D. MAT'L; RHc 24/30	H1150 COND.; AS REC'D. MAT'L	R_c 46 MIN; AS REC'D. MAT'L
INTERFACE MEMBERS DESCRIPTION – MATING MEMBERS	SURFACE FINISH	125 r m s	125 r m s	125 r m s	125 r m s
INTERFACE MEMBERS DESCRIPTION – MATING MEMBERS	SURFACE TREATMENT	PASSIVATE	ELECT. NICKEL PLATE; PASSIVATE	PASSIVATE	PASSIVATE
USING COMPONENT DESCRIPTION	PROCUREMENT DOCUMENT(S) OR P/N(S)	ME284 – 0358 – 0001	ME284 – 0358 – 0001	ME284 – 0161 – 0008	20M32015
USING COMPONENT DESCRIPTION	COMPONENT NAME	8 – INCH PROPELLANT SHUTOFF VALVE (PREVALVE)	8 – INCH PROPELLANT SHUTOFF VALVE (PREVALVE)	3 – INCH GOX PRESSURE REGULATOR	1 – 1 / 2 INCH HELIUM PRESSURE REGULATOR
USING COMPONENT DESCRIPTION	MANUFACTURER	PARKER – HANNIFIN	PARKER – HANNIFIN	PARKER – HANNIFIN	PARKER – HANNIFIN
USING COMPONENT DESCRIPTION	MANUFACTURER'S PART NUMBER	5670024	5670024	5640012M7	5640061
USING COMPONENT DESCRIPTION	USING VEHICLE(S) AND STAGE(S)	SATURN S – II	SATURN S – II	SATURN S – II	SATURN S – IC (QUALIFIED BACKUP)
USING COMPONENT DESCRIPTION	USING SYSTEM(S)	LOX & LH_2 FEED	LOX & LH_2 FEED	PREVIOUSLY IN LOX TANK PRESSURIZATION	DNA

Interface Category Number 10 (Sheet 2 of 3)

ROTATING SHAFT BALL AND ROLLER BEARINGS AND BALL SCREWS - Radial Bearing

		DESIGN CONCEPT NUMBER	10-5	10-6	10-7	10-8
COMPONENT DATA BOOK		CONTROL NUMBER(S)	S – IC / 207 – 5	S – II / 98 – 5	S – IC / 207 – 5	S – IVB / 210 – 3
		INTERFACE DASH NUMBER(S)	– 4	– 13 & – 13A	– 3	– 16, 16A, – 17, & – 17A
		WEAR ANALYSIS RATING (AVG.)	**8.3**	**8.4**	**8.9**	**8.9**
		PHOTOGRAPHS, PAGES	B – 4, – 5, & – 16	B – 3, & – 4	B – 4, – 5, – 6, – 7, – 14, & – 15	B – 7, & – 18
INTERFACE DESCRIPTION		INTERFACE PICTORIAL DESCRIPTION	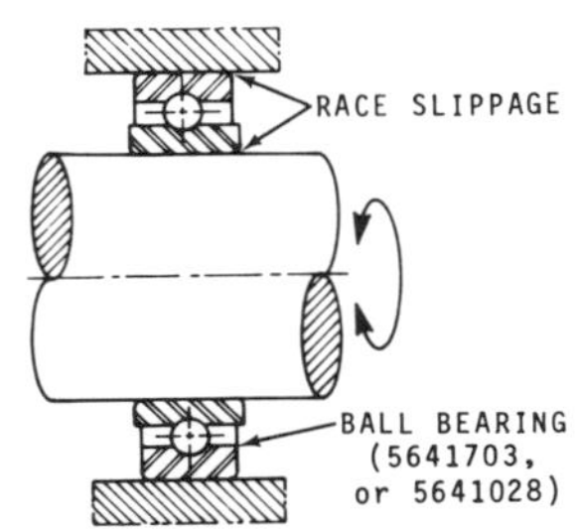		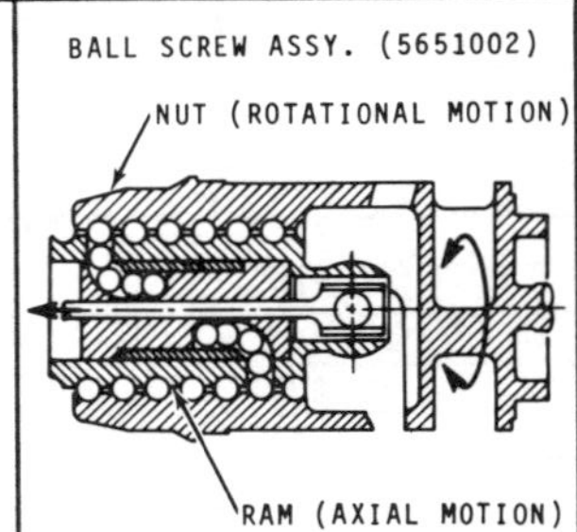	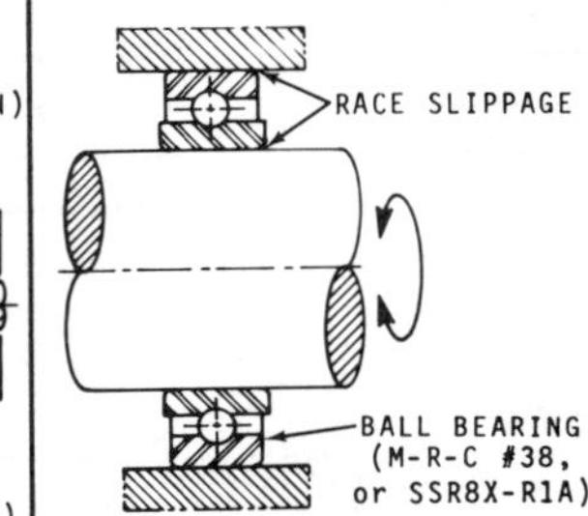
		BEARING TYPE	BALL	BALL	BALL SCREW	BALL
		OUTER RACE O. D., IN.	3.6772 + .0000, – .0006	2.0000 + .0000, – .0005	2.6756 – 2.6766; 1.9985 – 1.9995	.8567 – .8661; 1.1247 – 1.1250
		INNER RACE I. D., IN.	2.6766 + .0005, – .0000	1.0000 + .0000 – .0005	DNA (FIXED)	.3147 – .3150; .4997 – .5000
		OUTER RACE DIA. CLEARANCE, IN.	.0028 – .0044	.0020 – .0032; .0005 – .0020	.0000 – .0015	.0000 – .0006; – .0001 to + .0007
		INNER RACE DIA. CLEARANCE, IN.	.0000 – .0015	.0005 – 0015	DNA (FIXED)	.0003 – .0009; .0003 – .0009
		DESIGN OPERATING MEDIA	GHe, GN_2, or AIR	GOX	GHe, GN_2, or AIR	GHe
		DESIGN OPERATING TEMP. RANGE, °F	– 400 to + 165	– 180 to + 860	– 400 to + 165	– 423 to + 160
		DESIGN OPERATING LIFE, CYCLES	500	2000	500	200
		SPECIMEN OPERATION, CYCLES	500	147	500	30
INTERFACE MEMBERS DESCRIPTION	BEARING	PART NUMBER	5641703	5641028	5651002	M – R – C # 38 *; SSR8XR1A * *
		PART NAME	BEARING – ANGULAR CONTACT	BEARING – RADIAL, UPPER	BALL SCREW ASSY	BEARING
		MANUFACTURER P/N	INDUSTRIAL TECHTRONICS P / N 8653	INDUATRIAL TECHTRONICS P / N 7561	BEAVER PRECISION PROD. P / N B – 9477	MARLIN –ROCKWELL* ; NEW DEPARTURE **
		RACES MATERIAL	QQ – S – 763 440C CRES	QQ – S 763 440C CRES	PH13 – 8Mo	52100 STEEL; 440C CRES
		HEAT TREATMENT	R_c 55 MIN.	R_c 55 MIN.	R_c 46 MIN.	UNKNOWN; R_c 52–60
		SURFACE FINISH	125 r m s	125 r m s	125 r m s	125 r m s
		SURFACE TREATMENT	CHROME PLATE	ELECTROLIZE	PASSIVATE	UNKNOWN
	MATING MEMBERS	PART NUMBER	5651002; 5651013	5641020; 5641020; 5641014; 5641006	5641672; 5641703	511 – 46; 511 – 15
		PART NAME	BALL SCREW ASSY; HOUSING	CRANK; SLEEVE; BODY	BEARINGS	SHAFT; BODY
		MATERIAL	PH13 – 8 Mo; 303 CRES	303 Se; 304L; 17 – 4 PH	QQ – S – 763 440C CRES	INCONEL X – 750; 356 – T6 AL. ALY.
		HEAT TREATMENT	R_c 46 MIN; AS REC'D. MAT'L.	COND. B; COND. A; H1150 COND.	R_c 55 MIN.	AGE; COND. – T6
		SURFACE FINISH	125 r m s	125 r m s	125 r m s	125 r m s
		SURFACE TREATMENT	PASSIVATE	PASSIVATE	CHROME PLATE	PASSIVATE; ANODIZE
USING COMPONENT DESCRIPTION		PROCUREMENT DOCUMENT(S) OR P/N(S)	20M32015	ME284 – 0161 – 0008	20M32015	1A49988
		COMPONENT NAME	1 – 1 / 2 INCH HELIUM PRESSURE REGULATOR	3 – INCH GOX PRESSURE REGULATOR	1 – 1 / 2 INCH HELIUM PRESSURE REGULATOR	6 – INCH DIRECTIONAL CONTROL FUEL VENT VALVE
		MANUFACTURER	PARKER – HANNIFIN	PARKER – HANNIFIN	PARKER – HANNIFIN	AMETEK / CALMEC
		MANUFACTURER'S PART NUMBER	5640061	5640012M7	5640061	511 – 509
		USING VEHICLE(S) AND STAGE(S)	SATURN S – IC (QUALIFIED BACKUP)	SATURN S – II	SATURN S – IC (QUALIFIED BACKUP)	SATURN S – IVB
		USING SYSTEM(S)	DNA	PREVIOUSLY IN LOX TANK PRESSURIZATION	DNA	LH_2 TANK VENT

Interface Category Number 10
(Sheet 3 of 3)

ROTATING SHAFT BALL AND ROLLER BEARINGS AND BALL SCREWS - Radial Bearing

		DESIGN CONCEPT NUMBER	10-9	10-10		
COMPONENT DATA BOOK		CONTROL NUMBER(S)	S - IC / 206 - 1	J - 2 / 208 - 1		
		INTERFACE DASH NUMBER(S)	- 13, - 14, & - 15	- 5, & - 6		
		WEAR ANALYSIS RATING (AVG.)	**8.9**	**9.3**		
		PHOTOGRAPHS, PAGES	B - 3; - 12, & - 14	B - 2, - 3, - 4, - 11, - 12, & - 13		
INTERFACE DESCRIPTION		INTERFACE PICTORIAL DESCRIPTION	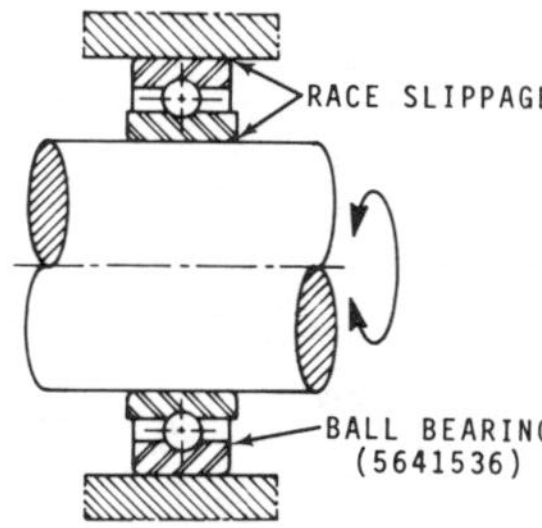	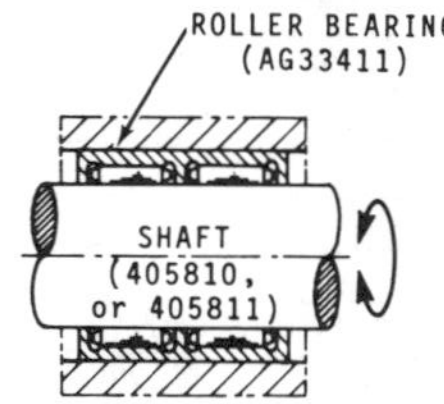		
		BEARING TYPE	BALL	ROLLER		
		OUTER RACE O. D., IN.	1.0625 + .0000, - .0010	1.4995 - 1.5000		
		INNER RACE I. D., IN.	.6250 ± .0007	1.0005 - 1.0021		
		OUTER RACE DIA. CLEARANCE, IN.	.0015 - .0030; .0012 - .0037	.0030 - .0039		
		INNER RACE DIA. CLEARANCE, IN.	+ .0005 to - .0011; + .0022 to - .0002	.0005 - .0026		
		DESIGN OPERATING MEDIA	LH_2	GH_2		
		DESIGN OPERATING TEMP. RANGE, °F	- 423 to + 160	- 420 to + 140		
		DESIGN OPERATING LIFE, CYCLES	2000	1308		
		SPECIMEN OPERATION, CYCLES	3318	292		
INTERFACE MEMBERS DESCRIPTION	BEARING	PART NUMBER	5641536	AG33411		
		PART NAME	BEARING	BRG, IDLER SHAFT; BRG. DRIVE SHAFT		
		MANUFACTURER P/N	FAFNIR P / N A538 - E9407	TORRINGTON CO. P / N AG33411		
		RACES MATERIAL	AISI 440C CRES	AMS5630 440C CRES		
		HEAT TREATMENT	UNKNOWN	R_c 53 MIN.		
		SURFACE FINISH	125 r m s	125 r m s		
		SURFACE TREATMENT	UNKNOWN	PASSIVATE		
	MATING MEMBERS	PART NUMBER	5641540; 5641532	405810; 405811		
		PART NAME	CRANK SHAFT; BODY	IDLER & DRIVE SHAFTS		
		MATERIAL	A - 286 CRES; 2014 - T4 AL. ALY.	LB0160 - 151 AISI TYPE 440C CRES		
		HEAT TREATMENT	R_c 27 - 37; - T6 COND.	R_c 53 - 60		
		SURFACE FINISH	125 r m s	8 r m s		
		SURFACE TREATMENT	PASSIVATE; ANODIZE	NONE		
USING COMPONENT DESCRIPTION		PROCUREMENT DOCUMENT(S) OR P/N(S)	20M32014	409920X7		
		COMPONENT NAME	2 - INCH NORMALLY CLOSED SUBMERGED LH_2 SHUTOFF VALVE	4 - INCH NORMALLY CLOSED MAIN FUEL VALVE		
		MANUFACTURER	PARKER - HANNIFIN	ROCKWELL / RKD		
		MANUFACTURER'S PART NUMBER	5640027	409920X7		
		USING VEHICLE(S) AND STAGE(S0	SATURN S - IC (QUALIFIED BACKUP)	SATURN S - II & SATURN S - IVB		
		USING SYSTEM(S)	DNA	J - 2 ENGINE LH_2 FEED		

Interface Category Number 11
(Sheet 1 of 4)

ROTATING PIN JOINT BEARINGS – Radial Bearing

		DESIGN CONCEPT NUMBER	11-1	11-2	11-3	11-4
COMPONENT DATA BOOK		CONTROL NUMBER(S)	S – IC / 206 – 1	S – II / 98 – 5	S – IC / 206 – 1	S – II / 98 – 5
		INTERFACE DASH NUMBER(S)	– 2 & – 3	– 5	– 6 & – 7	– 17
		WEAR ANALYSIS RATING (AVG.)	**1.0**	**6.8**	**7.0**	**7.0**
		PHOTOGRAPHS, PAGES	B – 4, – 6, & – 7	B – 2, & – 12	B – 4, – 6, & – 8	B – 2
INTERFACE DESCRIPTION		INTERFACE PICTORIAL DESCRIPTION	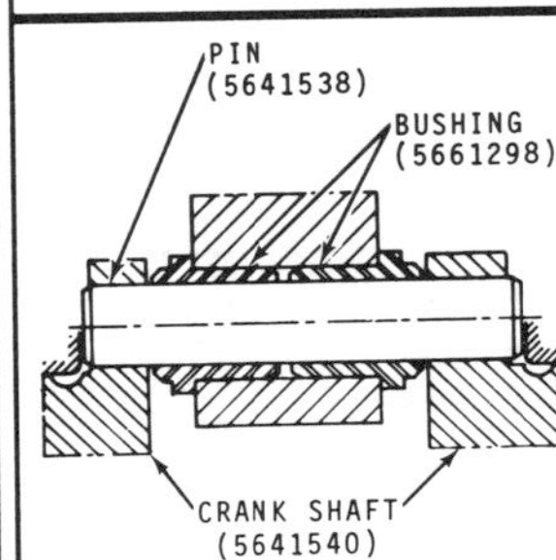	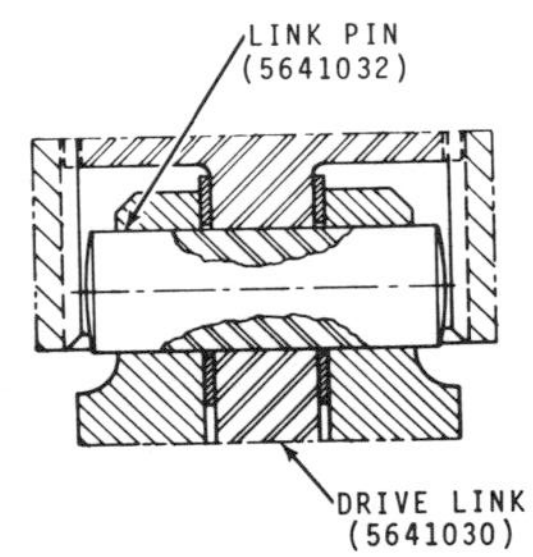	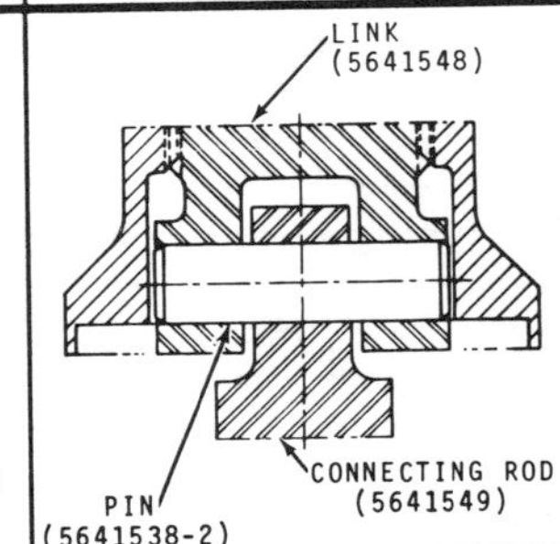	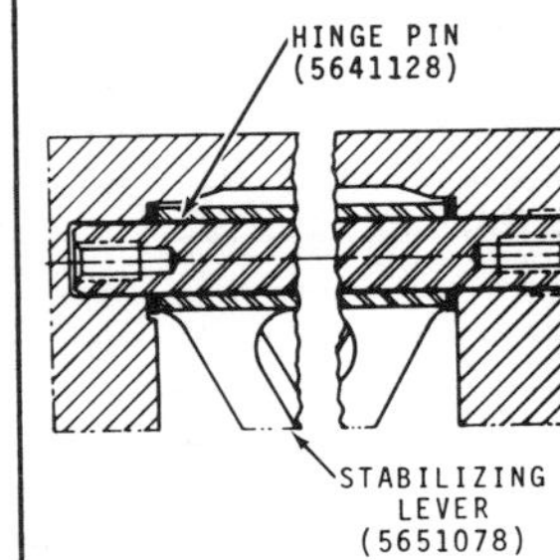
		BEARING TYPE	JOURNAL	JOURNAL	JOURNAL	JOURNAL
		BASIC BEARING DIA., IN.	.2185	.3135	.2182	.3740
		BEARING LENGTH, IN.	.728; .160 & .220	.270; .230 (2)	.280; .250 (2)	2.400; .420 & .680
		L / D	3.330; .732 & 1.010	.862; .735 (2)	1.282; 1.145 (2)	6.420; 1.120 & 1.820
		DIAMETRAL CLEARANCE, IN.	.0015 – .0025; .0002 – .0022	.0005 – .0020; .0005 – .0015	.0003 – .0015	.0010 – .0025
		DESIGN OPERATING MEDIA	LH_2	GOX	LH_2	GOX
		DESIGN OPERATING TEMP. RANGE, °F	– 423 to + 165	– 180 to + 860	– 423 to + 165	– 180 to + 860
		DESIGN OPERATING LIFE, CYCLES	2000	2000	2000	2000
		SPECIMEN OPERATION, CYCLES	3318	147	3318	147
INTERFACE MEMBERS DESCRIPTION	PIN OR BEARING	PART NUMBER	5641538 – 3	5641032	5641538 – 2	5641128
		PART NAME	PIN	PIN – LINK	PIN	PIN – HINGE
		MATERIAL	QQ – S – 763 440C CRES	AMS 5737 A – 286 CRES	QQ – S – 763 440C CRES	AMS 5737 A – 286 CRES
		HEAT TREATMENT	R_c 45 MIN.	AS REC'D. MAT'L. COND.	R_c 45 MIN.	AS REC'D. MAT'L. COND.
		SURFACE FINISH	125 r m s	32 r m s	125 r m s	32 r m s
		SURFACE TREATMENT	PASSIVATE	EVERLUBE 811	PASSIVATE	EVERLUBE 811
		OTHER SURFACE TREATMENTS	NONE	NONE	NONE	NONE
	MATING MEMBER(S)	PART NUMBER(S)	5661298; 5641540	5641030	5641549; 5641548	5651078
		PART NAME(S)	BUSHING; CRANK SHAFT	LINK – DRIVE	CONNECTING ROD; LINK	LEVER – STABILIZING
		MATERIAL(S)	BRONZE; A – 286 CRES	MIL – S – 7720 303 Se CRES	QQ – A – 225 / 6 2024 – T351 AL. ALY.	AMS 5737 A – 286 CRES
		HEAT TREATMENT(S)	FULL HARD DRAWN; R_c 27 – 37	COND. B	-T351 COND.	AS REC'D. MAT'L. COND.
		SURFACE FINISH(ES)	125 r m s	125 r m s	125 r m s	32 r m s
		SURFACE TREATMENT(S)	NONE; PASSIVATE	PASSIVATE	MIL – A – 8625 ANODIZE	PASSIVATE
		OTHER SURFACE TREATMENT(S)	NONE	NONE	NONE	NONE
USING COMPONENT DESCRIPTION		PROCUREMENT DOCUMENT(S) OR P/N(S)	20M32014	ME284 – 0161 – 0008	20M32014	ME284 – 0161 – 0008
		COMPONENT NAME	2 – INCH NORMALLY CLOSED SUBMERGED LH_2 SHUTOFF VALVE	3 – INCH GOX PRESSURE REGULATOR	2 – INCH NORMALLY CLOSED SUBMERGED LH_2 SHUTOFF VALVE	3 – INCH GOX PRESSURE REGULATOR
		MANUFACTURER	PARKER – HANNIFIN	PARKER – HANNIFIN	PARKER – HANNIFIN	PARKER – HANNIFIN
		MANUFACTURER'S PART NUMBER	5640027	5640012M7	5640027	5640012M7
		USING VEHICLE(S) AND STAGE(S)	SATURN S – IC (QUALIFIED BACKUP)	SATURN S – II	SATURN S – IC (QUALIFIED BACKUP)	SATURN S – II
		USING SYSTEM(S)	DNA	PREVIOUSLY IN LOX TANK PRESSURIZATION	DNA	PREVIOUSLY IN LOX TANK PRESSURIZATION

Interface Category Number 11
(Sheet 2 of 4)

ROTATING PIN JOINT BEARINGS
– Radial Bearing

		DESIGN CONCEPT NUMBER	11-5	11-6	11-7	11-8
COMPONENT DATA BOOK		CONTROL NUMBER(S)	S – IC / 207 – 5	S – IC / 207 – 5	J – 2 / 208 – 1	S – II / 98 – 5
		INTERFACE DASH NUMBER(S)	– 12 & – 13	– 10 and – 11	– 7 & – 8	– 4
		WEAR ANALYSIS RATING (AVG.)	**7.3**	**7.4**	**7.4**	**8.3**
		PHOTOGRAPHS, PAGES	B – 6, & – 19	B – 6, & – 19	B – 5, – 8, – 15, & – 16	B – 2, – 4, & – 12
INTERFACE DESCRIPTION		INTERFACE PICTORIAL DESCRIPTION	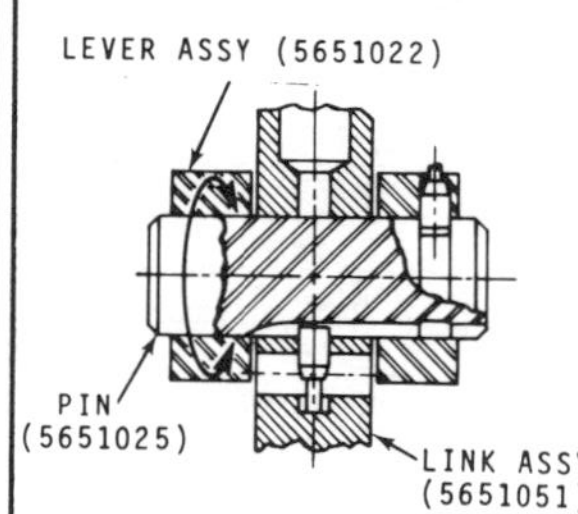	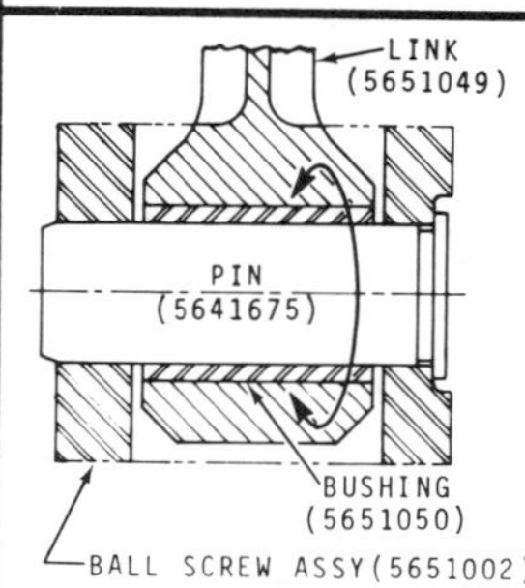	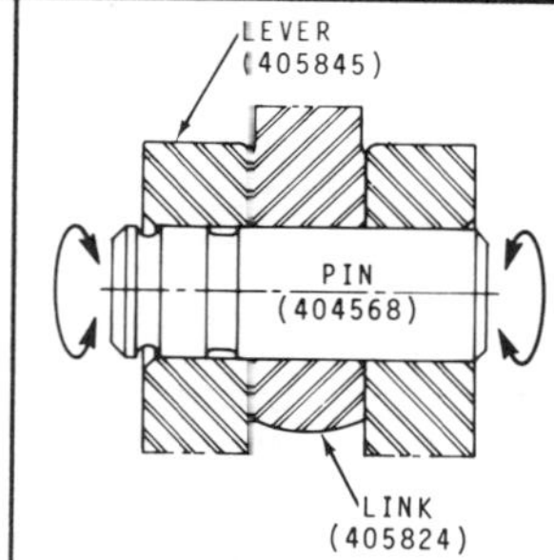	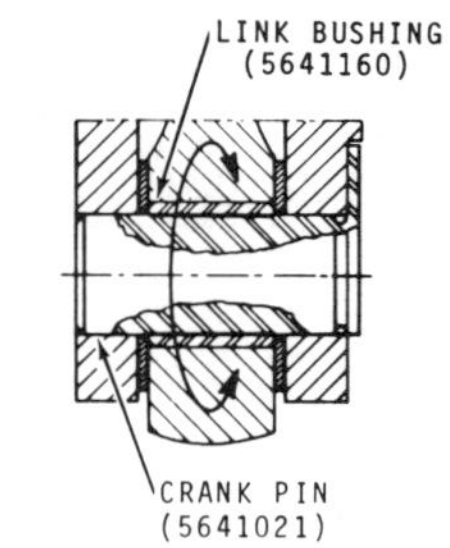
		BEARING TYPE	JOURNAL	JOURNAL	JOURNAL	JOURNAL
		BASIC BEARING DIA., IN.	.3120	.3750	.3750	.3125
		BEARING LENGTH, IN.	.308; .221 (2)	.625; .280 & .130	.311; .298 (2)	.340; .156 (2)
		L / D	.988; .710 (2)	1.670; .748 & .347	.830; .795 (2)	1.090; .500 (2)
		DIAMETRAL CLEARANCE, IN.	.0010 – .0025	.0008 – .0018	.0010 – .0018	.0005 – .0015
		DESIGN OPERATING MEDIA	GHe, GN_2, or AIR	GHe, GN_2 or AIR	GH_2 or AIR	GOX
		DESIGN OPERATING TEMP. RANGE, °F	– 400 to + 165	– 400 to + 165	– 420 to + 140	– 180 to + 860
		DESIGN OPERATING LIFE, CYCLES	500	500	1308	2000
		SPECIMEN OPERATION, CYCLES	500	500	292	147
INTERFACE MEMBERS DESCRIPTION	PIN OR BEARING	PART NUMBER	5651025	5641675	404568	5641021
		PART NAME	PIN – PIVOT	PIN – UPPER, POW. BEL. CONNECT.	PIN, LEVER – MPV	PIN – CRANK
		MATERIAL	AMS 5737 A – 286 CRES	AMS 5737 A – 286 CRES	QQ – S – 763 440C CRES	AMS 5737 A – 286 CRES
		HEAT TREATMENT	AS REC'D. MAT'L. COND.	AS REC'D. MAT'L. COND.	COND. A (R_c 53–60)	AS REC'D. MAT'L. COND.
		SURFACE FINISH	125 r m s	32 r m s	8 r m s	125 r m s
		SURFACE TREATMENT	EVERLUBE 811 – B	EVERLUBE 811	DRY FILM LUBRICANT	PASSIVATE
		OTHER SURFACE TREATMENTS	NONE	NONE	NONE	NONE
	MATING MEMBER(S)	PART NUMBER(S)	5651049; 5651022	5651050; 5651002	405824; 405845	5641160
		PART NAME(S)	LINK – DRIVE; LEVER ASSY.	BUSHING – LINK; BALL SCREW ASSY.	LINK, MPV; LEVER, MPV	BUSHING – LINK
		MATERIAL(S)	A – 286 CRES; 303 Se CRES	BEARIUM B – 11; PH13 – 8 Mo	MIL – S – 18732 431 CRES	QQ – L – 225 # 18 BEARIUM B – 10 BRONZE
		HEAT TREATMENT(S)	AS REC'D. MAT'L. COND.	AS REC'D. MAT'L. ; R_c 46 MIN.	180,000 to 200,000 PSI	AS REC'D. MAT'L. COND.
		SURFACE FINISH(ES)	125 r m s; 32 r m s	125 r m s	32 r m s; 63 r m s	125 r m s
		SURFACE TREATMENT(S)	PASSIVATE	NONE; PASSIVATE	NONE	NONE
		OTHER SURFACE TREATMENT(S)	NONE	NONE	NONE	NONE
USING COMPONENT DESCRIPTION		PROCUREMENT DOCUMENT(S) OR P/N(S)	20M32015	20M32015	409920M7	ME284 – 0161 – 0008
		COMPONENT NAME	1 – 1 / 2 INCH HELIUM PRESSURE REGULATOR	1 – 1 / 2 INCH HELIUM PRESSURE REGULATOR	4 – INCH NORMALLY CLOSED MAIN FUEL VALVE	3 – INCH GOX PRESSURE REGULATOR
		MANUFACTURER	PARKER – HANNIFIN	PARKER – HANNIFIN	ROCKWELL / RKD	PARKER – HANNIFIN
		MANUFACTURER'S PART NUMBER	5640061	5640061	409920M7	5640012M7
		USING VEHICLE(S) AND STAGE(S)	SATURN S – IC (QUALIFIED BACKUP)	SATURN S – IC (QUALIFIED BACKUP)	SATURN S – II 7 SATURN S – IVB	SATURN S – II
		USING SYSTEM(S)	DNA	DNA	J – 2 ENGINE LH_2 FEED	PREVIOUSLY IN LOX TANK PRESSURIZATION

Interface Category Number 11
(Sheet 3 of 4)

ROTATING PIN JOINT BEARINGS
- Radial Bearing

			11-9	11-10	11-10	11-11
		DESIGN CONCEPT NUMBER	11-9	11-10		11-11
COMPONENT DATA BOOK		CONTROL NUMBER(S)	S – II / 89 – 1	S – II / 89 – 1	S – II / 89 – 1	J – 2 / 208 – 1
		INTERFACE DASH NUMBER(S)	– 7 & – 8	– 11 & – 11A	– 9 & – 10	– 10 & – 12
		WEAR ANALYSIS RATING (AVG.)	**8.2**	**8.7**		**8.7**
		PHOTOGRAPHS, PAGES	B – 3, & – 10	B – 3, –4, – 10, & – 11	B – 3, – 4, – 10, & – 11	B – 5, – 6, & – 16
INTERFACE DESCRIPTION		INTERFACE PICTORIAL DESCRIPTION	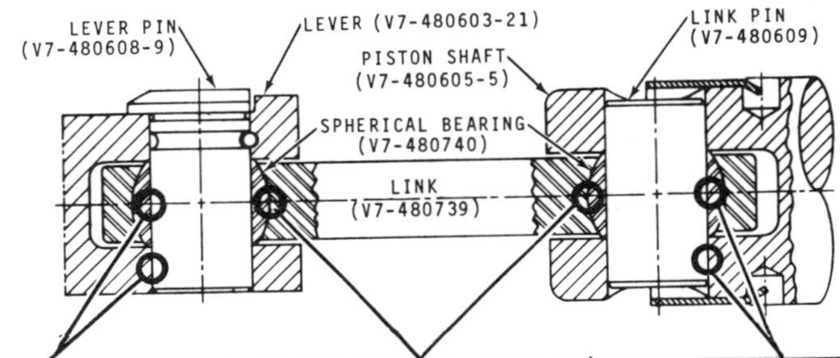			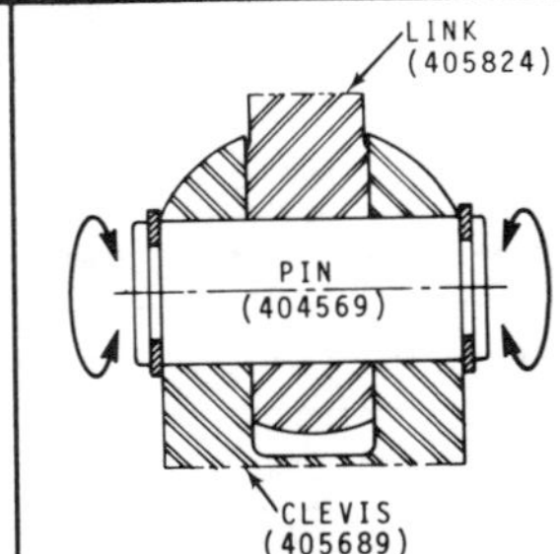
		BEARING TYPE	JOURNAL	SPHERICAL	JOURNAL	JOURNAL
		BASIC BEARING DIA., IN.	.4196	.5601	.4196	.3750
		BEARING LENGTH, IN.	.344; .200 (2)	.344	.344; .200 (2)	.311; .243 (2)
		L / D	.820; .477 (2)	.613	.820; .477 (2)	.830; .648 (2)
		DIAMETRAL CLEARANCE, IN.	.0004 – .0010; .0004 – .0008	.0001 – .0007	.0004 – .0010; .0004 – .0008	.0010 – .0018
		DESIGN OPERATING MEDIA	AIR	AIR	AIR	AIR
		DESIGN OPERATING TEMP. RANGE, °F	– 423 to + 160	– 423 to + 160	– 423 to + 160	– 420 to + 140
		DESIGN OPERATING LIFE, CYCLES	750	750	750	1308
		SPECIMEN OPERATION, CYCLES	591	591	591	292
INTERFACE MEMBERS DESCRIPTION	PIN OR BEARING	PART NUMBER	V7 – 480608 – 9	V7 – 480740	V7 – 480609	404569
		PART NAME	PIN, LEVER – FILL VALVE	BEARING, SPHERICAL	PIN, LINK – PREVALVE	PIN, CLEVIS, MPV
		MATERIAL	AMS 5667 INCONEL X – 750	QQ – C – 530 BERYLLIUM COPPER	AMS 5698 INCONEL X – 750	QQ – S 763 440C CRES
		HEAT TREATMENT	165,000 PSI MIN.	185,000 PSI MIN.	165,000 PSI MIN.	R_c 53 – 60
		SURFACE FINISH	8 r m s	16 r m s	8 r m s	8 r m s
		SURFACE TREATMENT	THIN DENSE CHROME PLATE	SILVER PLATE	THIN DENSE CHROME PLATE	DRY FILM LUBRICANT
		OTHER SURFACE TREATMENTS	NONE	NONE	NONE	NONE
	MATING MEMBER(S)	PART NUMBER(S)	V7 – 480740; V7 – 480603 – 21	V7 – 480739	V7 – 480740; V7 – 480605 – 5	405824; 405689
		PART NAME(S)	BEARING; LEVER	LINK, ACTUATOR – PREVALVE	BEARING, PISTON SHAFT	LINK, MPV; CLEVIS, MPV
		MATERIAL(S)	QQ – C – 530 BeCu; INCONEL 718	MB0170–04; INCONEL 718	QQ – C – 530 BeCu; INCONEL 718	MIL – S – 18732 431 CRES; QQ – S – 763 431 CRES
		HEAT TREATMENT(S)	185,000 PSI MIN; 175,000 PSI MIN.	175,000 PSI MIN.	185,000 PSI MIN; 175,000 PSI MIN	180,000 – 200,000 PSI
		SURFACE FINISH(ES)	16 r m s; 125 r m s (REAMED)	8 r m s	16 r m s	32 r m s
		SURFACE TREATMENT(S)	SILVER PLATE; PASSIVATE	PASSIVATE	SILVER PLATE, PASSIVATE	NONE
		OTHER SURFACE TREATMENT(S)	NONE	NONE	NONE	NONE
USING COMPONENT DESCRIPTION		PROCUREMENT DOCUMENT(S) OR P/N(S)	V7 – 480450 – 121	V7 – 480450 – 121	V7 – 480450 – 121	409920M7
		COMPONENT NAME	8 – INCH PROPELLANT FILL & DRAIN VALVE	8 – INCH PROPELLANT FILL & DRAIN VALVE	8 – INCH PROPELLANT FILL & DRAIN VALVE	4 – INCH NORMALLY CLOSED MAIN FUEL VALVE
		MANUFACTURER	ROCKWELL / LAD	ROCKWELL / LAD	ROCKWELL / LAD	ROCKWELL / RKD
		MANUFACTURER'S PART NUMBER	V7 – 480450 – 121	V7 – 480450 – 121	V7 – 480450 – 121	409920M7
		USING VEHICLE(S) AND STAGE(S)	SATURN S – II	SATURN S – II	SATURN S – II	SATURN S – II & SATURN S – IVB
		USING SYSTEM(S)	LH_2 TANK FILL & DRAIN	LH_2 TANK FILL & DRAIN	LH_2 TANK FILL & DRAIN	J – 2 ENGINE LH_2 FEED

Interface Category Number 11
(Sheet 4 of 4)

ROTATING PIN JOINT BEARINGS
- Radial Bearing

		DESIGN CONCEPT NUMBER	11-12	11-13	11-14	
COMPONENT DATA BOOK		CONTROL NUMBER(S)	S - IC / 207 - 5	S - IVB / 210 - 3	S - IC / 207 - 5	
		INTERFACE DASH NUMBER(S)	- 7, - 7A, - 8, & - 9	- 13, - 14, & - 15	- 14, & - 15	
		WEAR ANALYSIS RATING (AVG.)	**8.9**	**8.9**	**9.9**	
		PHOTOGRAPHS, PAGES	B - 5, - 17, & - 18	B - 6, - 7, & - 17	B - 6 & - 19	
INTERFACE DESCRIPTION		INTERFACE PICTORIAL DESCRIPTION	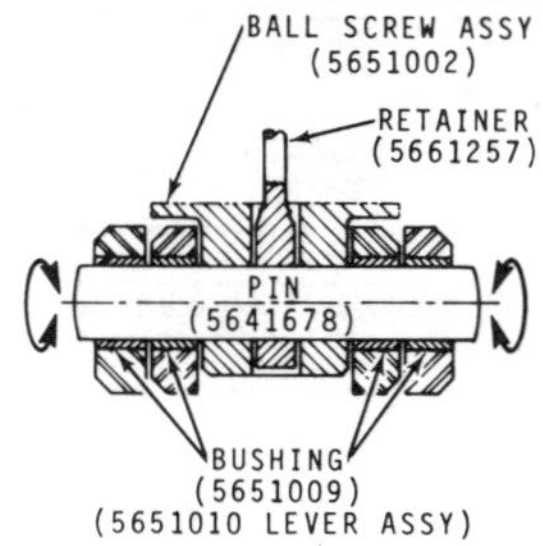	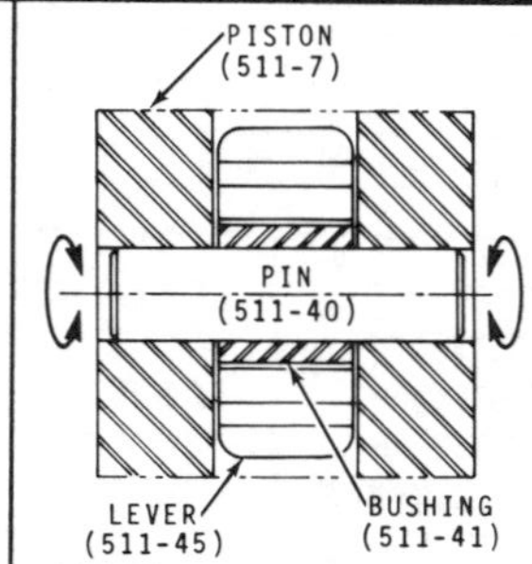	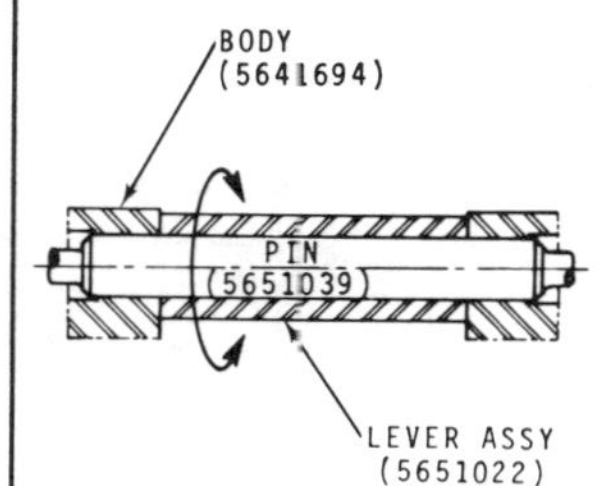	
		BEARING TYPE	JOURNAL	JOURNAL	JOURNAL	
		BASIC BEARING DIA., IN.	.3750	.2500; .3750	.3̇25	
		BEARING LENGTH, IN.	.260 (4); .252 (2); .210	.295 (2), .375; .375	1.592; 400 (2)	
		L / D	.693 (4); .672 (2); .561	1.180, 1.500; 1.000	5.100; 1.280 (2)	
		DIAMETRAL CLEARANCE, IN.	.0005 - .0020; .0005 - .0015; .0115 - .0140	.0003 - .0008; .0003 - .0007; .0002 - .0008	.0010 - .0030; .0005 - .0025	
		DESIGN OPERATING MEDIA	GHe, GN_2 or AIR	GH_2 or LH_2	GHe, GN_2 or AIR	
		DESIGN OPERATING TEMP. RANGE, °F	- 400 to + 165	- 400 to + 160	- 400 to + 165	
		DESIGN OPERATING LIFE, CYCLES	500	200	500	
		SPECIMEN OPERATION, CYCLES	500	30	500	
INTERFACE MEMBERS DESCRIPTION	PIN OR BEARING	PART NUMBER(S)	5641678	511 - 40; 511 - 41	5651039	
		PART NAME(S)	PIN, BALL SCREW, LEVER	PIN; BUSHING	PIN, PIVOT, POWER BELLOWS	
		MATERIAL	AMS 5737 A - 286 CRES	AMS 5643 17 - 4PH CRES	AMS 5737 A - 286 CRES	
		HEAT TREATMENT	AS REC'D. MAT'L. COND.	COND. H - 925	AS REC'D. MAT'L. COND	
		SURFACE FINISH	32 r m s	16 r m s	32 r m s	
		SURFACE TREATMENT	EVERLUBE 811	PASSIVATE; LUBELOCK # 4396	EVERLUBE 811	
		OTHER SURFACE TREATMENTS	NONE	NONE; PASSIVATE	NONE	
	MATING MEMBER(S)	PART NUMBER(S)	5651009; 5651002; 5661257	511 - 7, 511 - 41; 511 - 45	5651022; 5641694	
		PART NAME(S)	BUSHING; BALL SCREW; RETAINER	PISTON, BUSHING; LEVER	LEVER ASSY; BODY ASSY	
		MATERIAL(S)	BEARIUM B - 11; PH13 - 8 MO; A - 286 CRES	AMS 5643 17 - 4 PH CRES	303 Se CRES; 7075 - T73 AL. ALY.	
		HEAT TREATMENT(S)	AS REC'D; R_c 46 MIN.; AS REC'D	COND. H - 925	AS REC'D; COND. - T73	
		SURFACE FINISH(ES)	32 r m s; 32 r m s; 125 r m s	125 r m s; 16 r m s; 32 r m s	32 r m s; 125 r m s	
		SURFACE TREATMENT(S)	NONE; PASSIVATE, PASSIVATE	PASSIVATE, LUBELOCK #4396; LUBELOCK # 4396	PASSIVATE; ANODIZE	
		OTHER SURFACE TREATMENT(S)	NONE	NONE, PASSIVATE, PASSIVATE	NONE	
USING COMPONENT DESCRIPTION		PROCUREMENT DOCUMENT(S) OR P/N(S)	20M32015	1A49988	20M32015	
		COMPONENT NAME	1 - 1 / 2 INCH HELIUM PRESSURE REGULATOR	6 - INCH DIRECTIONAL CONTROL FUEL VENT VALVE	1 - 1 / 2 INCH HELIUM PRESSURE REGULATOR	
		MANUFACTURER	PARKER - HANNIFIN	AMETEK / CALMEC	PARKER - HANNIFIN	
		MANUFACTURER'S PART NUMBER	5640061	511 - 509	5640061	
		USING VEHICLE(S) AND STAGE(S)	SATURN S - IC (QUALIFIED BACKOU)	SATURN S - IVB	SATURN S - IC (QUALIFIED BACKUP)	
		USING SYSTEM(S)	DNA	LH_2 TANK VENT	DNA	

Interface Category Number 12 (Sheet 1 of 4)

AXIAL/ROTATING THRUST BEARINGS - Face Bearing

		DESIGN CONCEPT NUMBER	12-1	12-2	12-3	12-4
COMPONENT DATA BOOK		CONTROL NUMBER(S)	S-IC/206-1	S-IVB/210-3	S-II/98-5	S-II/90-1
		INTERFACE DASH NUMBER(S)	-1	-19, -19A, -20, & -20A	-16 & -16A	-11 & -12
		WEAR ANALYSIS RATING (AVG.)	**6.0**	**6.6**	**6.7**	**7.2**
		PHOTOGRAPHS, PAGES	B-3, & -5	B-8, -20, & -21	B-3, -4, & -20	B-4, -17, & -18
INTERFACE DESCRIPTION		INTERFACE PICTORIAL DESCRIPTION	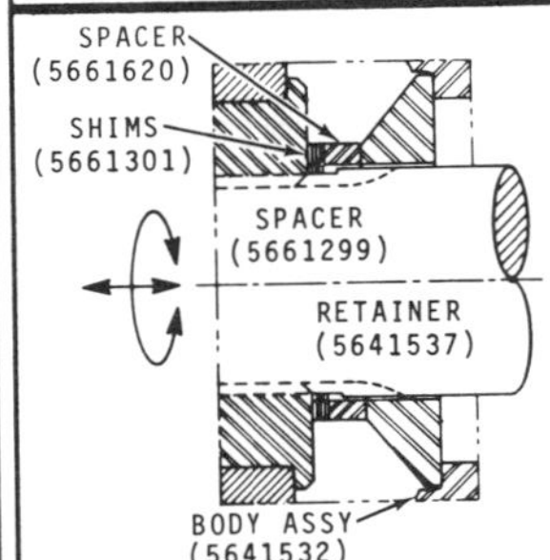	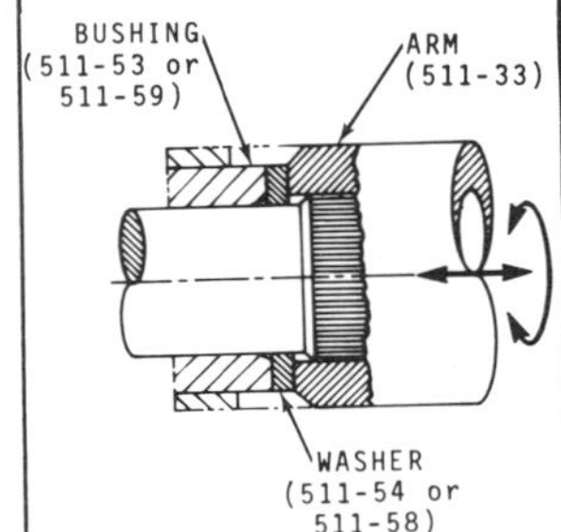	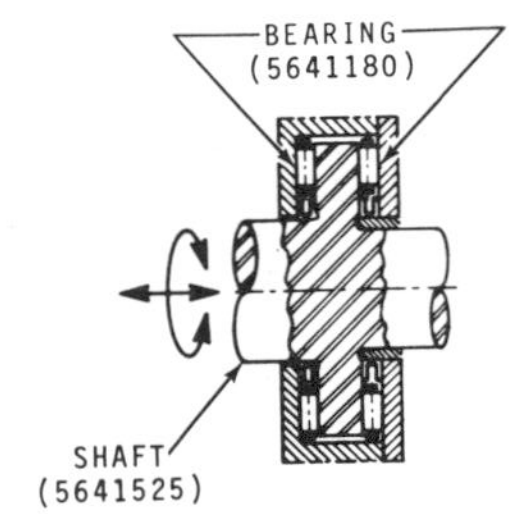	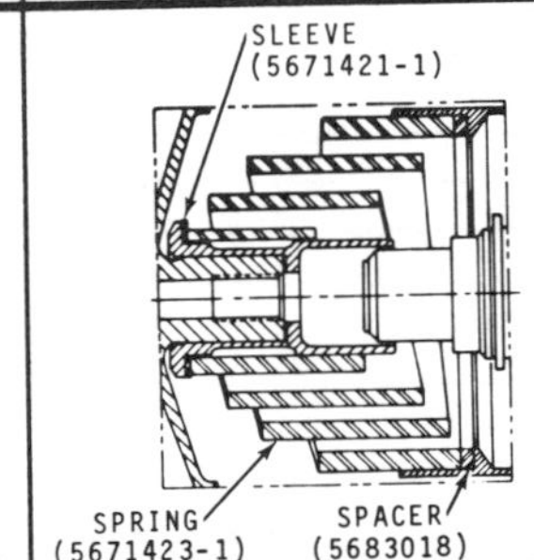
		BEARING TYPE	METAL-TO-METAL	METAL-TO-METAL	ROLLER BEARING	MEATL-TO-METAL
		BEARING FACE O. D., IN.	.750[1-4]; 1.12[1,5]	.625	1.500	1.810; 3.885
		BEARING FACE I. D., IN.	.640[1-4]; 1.00[1,5]	.515[1-2]; .440[3-4]; .460[2-5]; .408[4-5]	1.100	1.310; 3.400
		DESIGN OPERATING MEDIA	LH2	GH2 or LH2	GOX	LOX or LH2
		DESIGN OPERATING TEMP. RANGE, °F	-423 to +165	-423 to +160	-180 to +860	-423 to +160
		DESIGN OPERATING LIFE, CYCLES	2000	200	2000	1000
		SPECIMEN OPERATION, CYCLES	3318	30	147	1178
INTERFACE MEMBERS DESCRIPTION	BEARING MEMBER NO. 1	PART NUMBER(S)	5661299[1]; 5661620[2]; * * 5661301[3]; 5641537[4]	511-53[1]; 511-54[2]; 511-59[3]; 511-58[4]	5641180	5671423-1
		PART NAME(S)	SPACER[1,2]; SHIMS[3]; RETAINER[4]	BUSHING[1,3]; WASHER[2,4]	BEARING - THRUST	SPRING - VOLUTE, STROKE
		MATERIAL	6061 AL. ALY.[1,2]; 300 CRES[3]; A-286[4]	AMS5643 17-4PH[1,3]; MIL-7720 303[2,4]	AMS 5796 HAYNES ALLOY 25	AMS 5597 INCONEL 718
		HEAT TREATMENT	-T651[1,2]; AS REC'D[3]; R_C 27-37[4]	COND. H-925[1,3]; AS REC'D[2,4]	R_C 49 MIN.	AMS 5597
		SURFACE FINISH	125 r m s	125 r m s[1,3]; 63 r m s[2,4]	NOT SPECIFIED	125 r m s
		SURFACE TREATMENT	ANODIZE[1,2]; PASSIVATE[3,4]	PASSIVATE	NONE	LUBECO 905
		OTHER SURFACE TREATMENTS	NONE	NONE	NONE	PASSIVATE
	BEARING MEMBER NO. 2	PART NUMBER(S)	5641532[5]	511-33[5]	5641525	5671421-1; 5683018
		PART NAME(S)	BODY ASSY.	ARM - FLAPPER	SHAFT - BUTTERFLY	SLEEVE, SPACER
		MATERIAL	QQ-A-367 2014-T4 AL. ALY.	QQ-A-327 6061-T6 AL. ALY.	AMS 5737 A-286 CRES	A-286 CRES; 321 CRES
		HEAT TREATMENT	COND. - T6	COND. - T6	AS REC'D. MAT'L. COND.	R_C 24 MIN; AS REC'D. MAT'L.
		SURFACE FINISH	125 r m s	32 r m s	32 r m s	125 r m s
		SURFACE TREATMENT	MIL-A-8625 ANODIZE	MIL-A-8625 ANODIZE	PASSIVATE	PASSIVATE
		OTHER SURFACE TREATMENTS	NONE	NONE	NONE	NONE
USING COMPONENT DESCRIPTION		PROCUREMENT DOCUMENT(S) OR P/N(S)	20M32014	1A49988	ME284-0161-0008	ME284-0358-0001
		COMPONENT NAME	2-INCH NORMALLY CLOSED SUBMERGED LH_2 SHUTOFF VALVE	6-INCH DIRECTIONAL CONTROL FUEL VENT VALVE	3-INCH GOX PRESSURE REGULATOR	8-INCH PROPELLANT SHUTOFF VALVE (PREVALVE)
		MANUFACTURER	PARKER-HANNIFIN	AMETEK / CALMEC	PARKER-HANNIFIN	PARKER-HANNIFIN
		MANUFACTURER'S PART NUMBER	5640027	511-509	5640012M7	5670024
		USING VEHICLE(S) AND STAGE(S)	SATURN S-IC (QUALIFIED BACKUP)	SATURN-IVB	SATURN S-II	SATURN S-II
		USING SYSTEM(S)	DNA	LH_2 TANK VENT	PREVIOUSLY IN LOX TANK PRESSURIZATION	LOX & LH_2 FEED

Interface Category Number 12
(Sheet 2 of 4)

AXIAL/ROTATING THRUST BEARINGS
- Face Bearing

		DESIGN CONCEPT NUMBER	12-5	12-6	12-7	12-8
COMPONENT DATA BOOK		CONTROL NUMBER(S)	S – II / 89 – 1	S – IC / 206 – 1	J – 2 / 208 – 1	S – II / 89 – 1 ; S – II / 90 – 1
		INTERFACE DASH NUMBER(S)	– 17, – 17A, – 17B, & – 17C	– 4 & – 5	– 2 & – 2A	– 3 & – 3A; – 4 & – 4A
		WEAR ANALYSIS RATING (AVG.)	**8.3**	**8.9**	**9.1**	**9.1**
		PHOTOGRAPHS, PAGES	B – 5, – 6, & – 19	B – 3, & – 9	B – 2, – 3, – 4, & – 10	B – 2, & – 8; B – 2, & – 20
INTERFACE DESCRIPTION		INTERFACE PICTORIAL DESCRIPTION	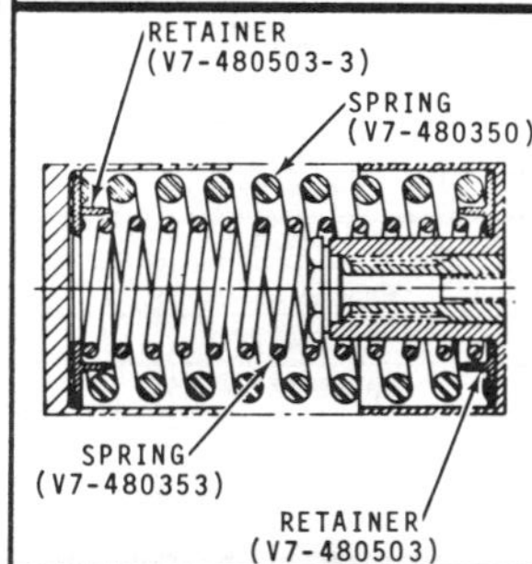	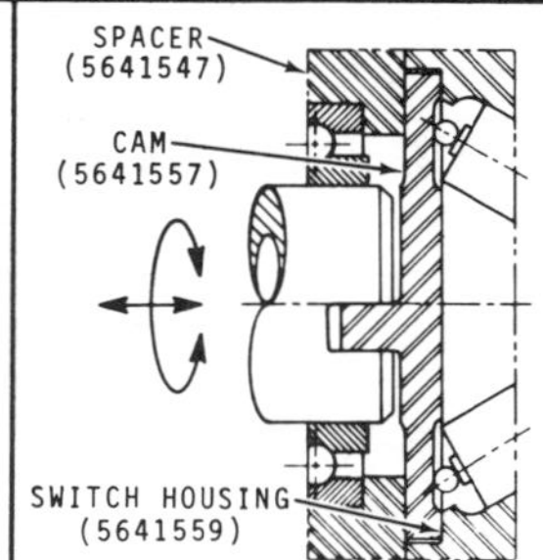	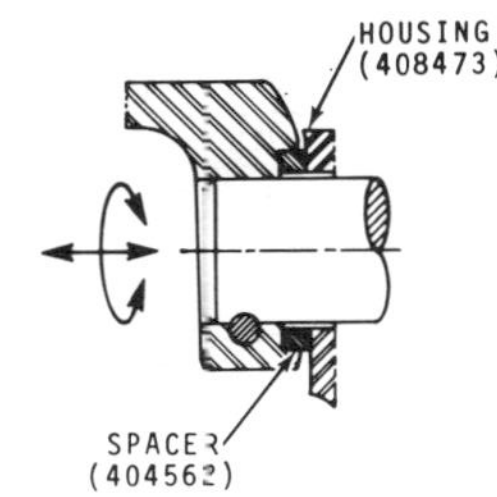	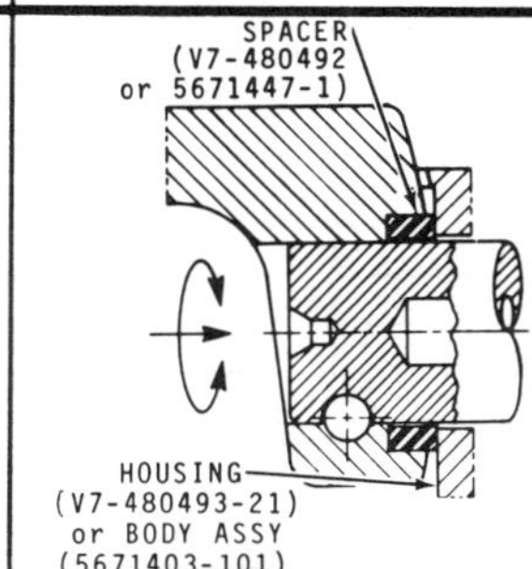
		BEARING TYPE	METAL – TO – METAL	METAL – TO – METAL	PLASTIC	PLASTIC
		BEARING FACE O. D., IN.	1.600; 2.500	1.225	1.373	1.310; 1.448
		BEARING FACE I. D., IN.	1.216; 1.876	.90; 1.140	1.062	1.031; 1.250
		DESIGN OPERATING MEDIA	GHe	LH_2	LH_2	LOX or LH_2
		DESIGN OPERATING TEMP. RANGE, °F	– 423 to + 160	– 423 to + 165	– 420 to + 140	– 423 to + 160
		DESIGN OPERATING LIFE, CYCLES	750	2000	1308	750; 1000
		SPECIMEN OPERATION, CYCLES	591	3318	292	591; 1178
INTERFACE MEMBERS DESCRIPTION	BEARING MEMBER NO. 1	PART NUMBER(S)	V7 – 480350; V7 – 480353	5641557	404562	V7 – 480492; 5671447 – 1
		PART NAME(S)	OUTER SPRING; INNER SPRING	CAM	SPACER – MPV	SPACER
		MATERIAL	MB0170 – 044 ELGILOY	QQ – A – 225 / 8 6061 – T651 AL. ALY.	25 % ASBESTOS FILLED TEFLON	25 % ASBESTOS FILLED TEFLON
		HEAT TREATMENT	275,000 PSI MIN. U. T. S.	COND. – T651	STRESS RELIEVE (RA0105 – 002)	STRESS RELIEVE (MA0105 – 047)
		SURFACE FINISH	125 r m s	125 r m s	125 r m s	125 r m s
		SURFACE TREATMENT	PASSIVATE	NAS1192 HARD ANODIZE	NONE	NONE
		OTHER SURFACE TREATMENTS	NONE	NONE	NONE	NONE
	BEARING MEMBER NO. 2	PART NUMBER(S)	V7 – 480503 – 3; V7 – 480503	5641547; 5641559	408473	V7 – 480493 – 21; 5671403 – 101
		PART NAME(S)	RETAINER – SPRING	SPACER; SWITCH HOUSING	HOUSING, MAIN FUEL VALVE ASSY.	HOUSING; BODY ASSY
		MATERIAL	QQ – S – 763 303 Se CRES	303 CRES; 304L CRES	(405806 – 3 ROUGH CASTING)	TENS 50 – T60; 6061 – T6
		HEAT TREATMENT	COND. A	AS REC'D. MAT'L. COND.	COND. – T6	COND. – T60; COND. – T6
		SURFACE FINISH	125 r m s	32 r m s; 125 r m s	63 r m s	63 r m s
		SURFACE TREATMENT	PASSIVATE	PASSIVATE	LB0125 – 103 HARD ANODIZE	ANODIZE
		OTHER SURFACE TREATMENTS	NONE	NONE	NONE	NONE
USING COMPONENT DESCRIPTION		PROCUREMENT DOCUMENT(S) OR P/N(S)	V7 – 480450 – 121	20M32014	409920X7	V7 – 480450 – 121; ME284 – 0358 – 0001
		COMPONENT NAME	8 – INCH PROPELLANT FILL & DRAIN VALVE	2 – INCH NORMALLY CLOSED SUBMERBED LH_2 SHUTOFF VALVE	4 – INCH NORMALLY CLOSED MAIN FUEL VALVE	8 – INCH FILL & DRAIN VALVE; 8 – INCH PREVALVE
		MANUFACTURER	ROCKWELL / LAD	PARKER – HANNIFIN	ROCKWELL / RKD	ROCKWELL / LAD; PARKER – HANNIFIN
		MANUFACTURER'S PART NUMBER	V7 – 480450 – 121	5640027	409920X7	V7 – 480450 – 121; 5640024
		USING VEHICLE(S) AND STAGE(S)	SATURN S – II	SATURN S – IC (QUALIFIED BACKUP)	SATURN S – II & SATURN S – IVB	SATURN S – II
		USING SYSTEM(S)	LOX & LH_2 TANK FILL & DRAIN	DNA	J – 2 ENGINE LH_2 FEED	LOX & LH_2 FILL & DRAIN; LOX & LH_2 FEED

Interface Category Number 12
(Sheet 3 of 4)

AXIAL/ROTATING THRUST BEARINGS
- Face Bearing

Section		DESIGN CONCEPT NUMBER	12-9	12-10	12-11	12-12
COMPONENT DATA BOOK		CONTROL NUMBER(S)	J-2/208-1	S-IVB/209-2	S-II/90-1	S-II/100-1
		INTERFACE DASH NUMBER(S)	-9 & -11	-10	-14	-2 & -2A
		WEAR ANALYSIS RATING (AVG.)	**9.2**	**9.5**	**9.5**	**9.5**
		PHOTOGRAPHS, PAGES	B-5, -15, & -16	B-5, & -15	B-7	81 & 86
INTERFACE DESCRIPTION		INTERFACE PICTORIAL DESCRIPTION	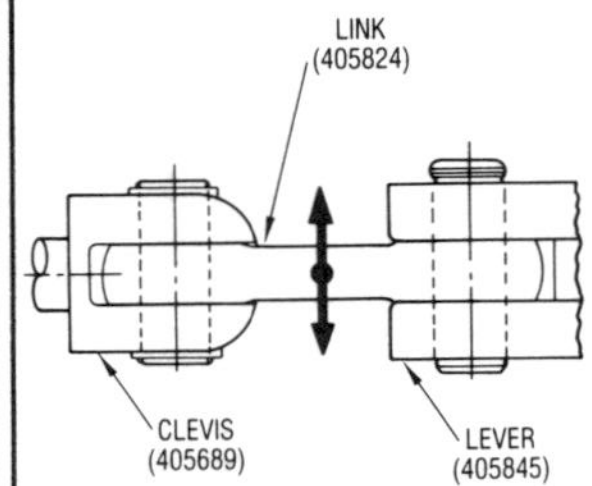	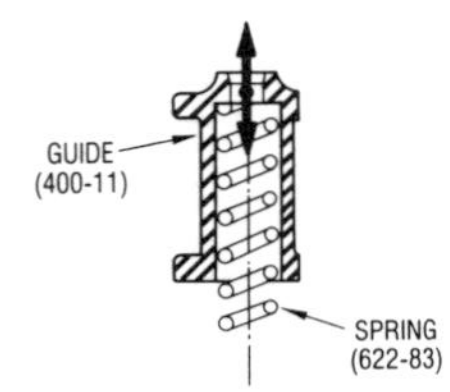	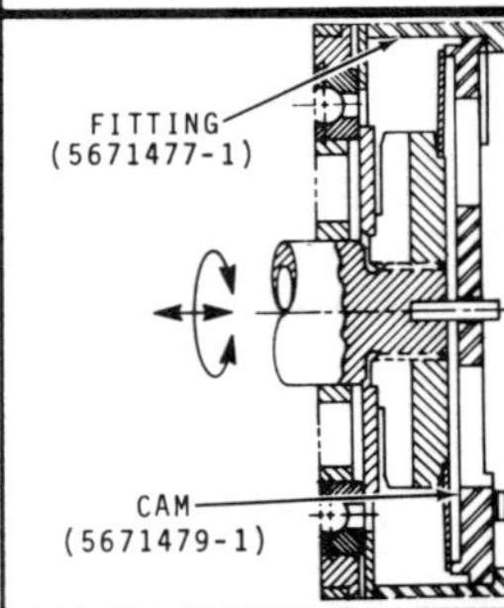	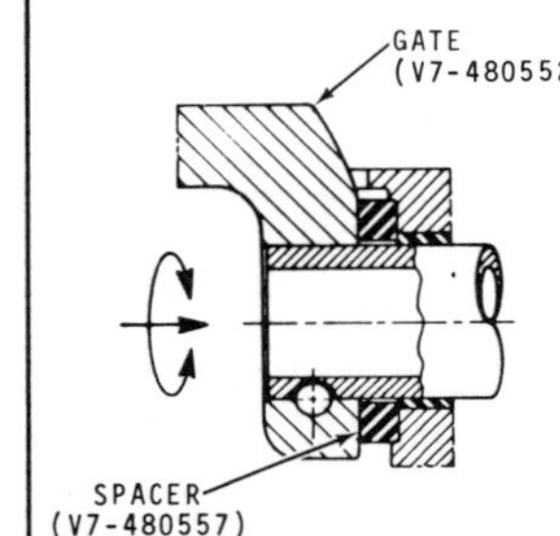
		BEARING TYPE	METAL-TO-METAL	METAL-TO-METAL	METAL-TO-METAL	PLASTIC
		BEARING FACE O. D., IN.	.750	.080	2.840	.673
		BEARING FACE I. D., IN.	.375	.060	2.720	.440
		DESIGN OPERATING MEDIA	AIR	GHe	LOX or LH_2	LOX or LH_2
		DESIGN OPERATING TEMP. RANGE, °F	-420 to +140	-423 to +160	-423 to +160	-423 to +160
		DESIGN OPERATING LIFE, CYCLES	1308	5000	1000	750
		SPECIMEN OPERATION, CYCLES	292	10,121	1178	935
INTERFACE MEMBERS DESCRIPTION	BEARING MEMBER NO. 1	PART NUMBER	405824	400-11	5671479-1	V7-480557
		PART NAME	LINK, MAIN PROP. VALVE	GUIDE, PILOT POPPET SOL. VALVE	CAM	SPACER-RECIRC. VALVE
		MATERIAL	MIL-S-18732 431 CRES	MIL-S-7720 18-8/303 CRES	QQ-A-225/9 7075-T6 AL. ALY.	25 % ASBESTOS FILLED TEFLON
		HEAT TREATMENT	180,000-200,000 PSI	COND. B	COND.-T6	STRESS RELIEVE (MA0105-047)
		SURFACE FINISH	32 r m s	125 r m s	16 r m s	125 r m s
		SURFACE TREATMENT	SOLID FILM LUBRICANT	PASSIVATE	LUBECO 905	NONE
		OTHER SURFACE TREATMENTS	NONE	NONE	MIL-A-8625 HARD ANODIZE	NONE
	BEARING MEMBER NO. 2	PART NUMBER(S)	405845; 405689	622-83	5671477-1	V7-480 0552
		PART NAME(S)	LEVER, MPV; CLEVIS, MPV	SPRING, PILOT BALL RETURN	FITTING - POSITION SWITCH	GATE - RECIRCULATION VALVE
		MATERIAL	431 CRES	QQ-W-423 18-8/302 CRES	QQ-S-763 321, 347 or 304L CRES	AMS 5737 A-286 CRES
		HEAT TREATMENT	180,000-200,000 PSI	AS REC'D. MAT'L. COND.	AS REC'D. MAT'L. COND.	AS REC MAT'L. COND.
		SURFACE FINISH	63 r m s; 32 r m s	125 r m s	16 r m s	125 r m s
		SURFACE TREATMENT	NONE	PASSIVATE	PASSIVATE	PASSIVATE
		OTHER SURFACE TREATMENTS	NONE	NONE	NONE	NONE
USING COMPONENT DESCRIPTION		PROCUREMENT DOCUMENT(S) OR P/N(S)	409920X7	1B43660-515	ME284-0358-0001	V7-480328-91
		COMPONENT NAME	4-INCH NORMALLY CLOSED MAIN FUEL VALVE	1/2-INCH NORMALLY CLOSED HELIUM SOLENOID VALVE	8-INCH PROPELLANT SHUTOFF VALVE (PREVALVE)	2-INCH NORMALLY OPEN RECIRCULATION VALVE
		MANUFACTURER	ROCKWELL/RKD	AMETEK/CALMEC	PARKER-HANNIFIN	ROCKWELL/LAD
		MANUFACTURER'S PART NUMBER	409920X7	548-515	5640024	V7-480328-91
		USING VEHICLE(S) AND STAGE(S)	SATURN S-II & SATURN S-IVB	SATURN S-IVB	SATURN S-II	SATURN S-II
		USING SYSTEM(S)	J-2 ENGINE LH_2 FEED	PRESSURE CONTROL	LOX & LH_2 FEED	LOX & LH_2 RECIRCULATION

Interface Category Number 12
(Sheet 4 of 4)

AXIAL/ROTATING THRUST BEARINGS
- Face Bearing

	DESIGN CONCEPT NUMBER	12-13	12-14		
COMPONENT DATA BOOK	CONTROL NUMBER(S)	S – IVB / 209 – 2	S – IVB / 209 – 2		
	INTERFACE DASH NUMBER(S)	– 6	– 11		
	WEAR ANALYSIS RATING (AVG.)	**9.6**	**10.0**		
	PHOTOGRAPHS, PAGES	B – 4, – 5, – 13, & – 14	B – 6, & – 16		
INTERFACE DESCRIPTION	INTERFACE PICTORIAL DESCRIPTION	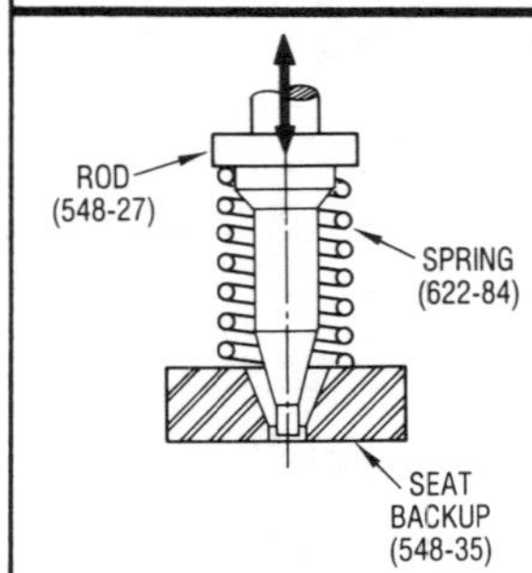	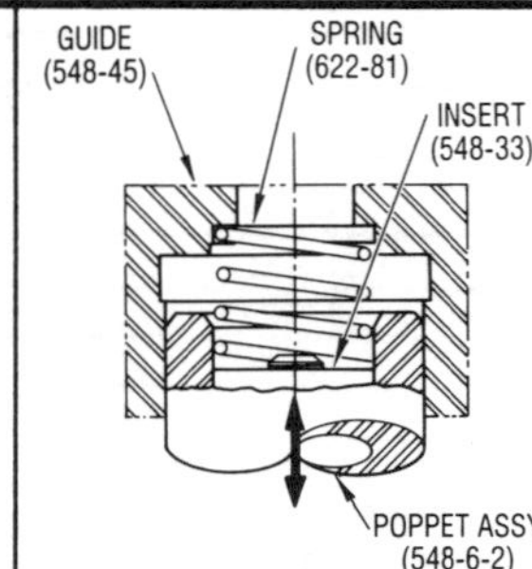		
	BEARING TYPE	METAL – TO – METAL	METAL – TO – METAL		
	BEARING FACE O. D., IN.	.183	.437		
	BEARING FACE I. D., IN.	.143	.385		
	DESIGN OPERATING MEDIA	GHe	GHe		
	DESIGN OPERATING TEMP. RANGE, °F	– 423 to + 160	– 423 to + 160		
	DESIGN OPERATING LIFE, CYCLES	5000	5000		
	SPECIMEN OPERATION, CYCLES	10,121	10,121		
INTERFACE MEMBERS DESCRIPTION – BEARING MEMBER NO. 1	PART NUMBER	622 – 84	622 – 81		
	PART NAME	SPRING, SOLENOID ROD RETURN	SPRING, POPPET RETURN		
	MATERIAL	QQ – W – 423 302 CRES	QQ – W – 423 302 CRES		
	HEAT TREATMENT	AS REC'D. MAT'L. COND.	AS REC'D. MAT'L. COND.		
	SURFACE FINISH	125 r m s	125 r m s		
	SURFACE TREATMENT	PASSIVATE	PASSIVATE		
	OTHER SURFACE TREATMENTS	NONE	NONE		
BEARING MEMBER NO. 2	PART NUMBER	548 – 35; 548 – 27	548 – 33		
	PART NAME	SEAT; ROD	INSERT, SOLENOID VALVE POPPET		
	MATERIAL	17 – 4PH CRES; 303 CRES	AMS5643 17 – 4PH CRES		
	HEAT TREATMENT	COND. H – 925; COND. B	AS REC'D. MAT'L. COND.		
	SURFACE FINISH	125 r m s	125 r m s		
	SURFACE TREATMENT	PASSIVATE	PASSIVATE		
	OTHER SURFACE TREATMENTS	NONE	NONE		
USING COMPONENT DESCRIPTION	PROCUREMENT DOCUMENT(S) OR P/N(S)	1B43660 – 515	1B43660 – 515		
	COMPONENT NAME	1 / 2 – INCH NORMALLY CLOSED HELIUM SOLENOID VALVE	1 / 2 – INCH NORMALLY CLOSED HELIUM SOLENOID VALVE		
	MANUFACTURER	AMETEK / CALMEC	AMETEK / CALMEC		
	MANUFACTURER'S PART NUMBER	548 – 515	548 – 515		
	USING VEHICLE(S) AND STAGE(S)	SATURN S – IVB	SATURN S – IVB		
	USING SYSTEM(S)	PRESSURE CONTROL	PRESSURE CONTROL		

Interface Category Number 13
(Sheet 1 of 3)

MISCELLANEOUS LOCATING/SUPPORT GUIDES/BEARINGS

	DESIGN CONCEPT NUMBER	13-1	13-2	13-3	13-4
COMPONENT DATA BOOK	CONTROL NUMBER(S)	S – II / 104 – 2; S – IC / 1 – 2	S – IC / 207 – 5	S – II / 96 – 5	S – II / 104 – 2; S – IC / 1 – 2
	INTERFACE DASH NUMBER(S)	– 2; – 7	– 6 & – 6A	– 7	– 7; – 1
	WEAR ANALYSIS RATING (AVG.)	**6.8**	**6.9**	**8.5**	**9.8**
	PHOTOGRAPHS, PAGES	B – 2, & – 5; B – 2, – 6, – 7, & – 8	B – 5, – 6, – 7, & – 17	B – 7	B – 3, & – 10; B – 3, & – 4
INTERFACE DESCRIPTION	INTERFACE PICTORIAL DESCRIPTION	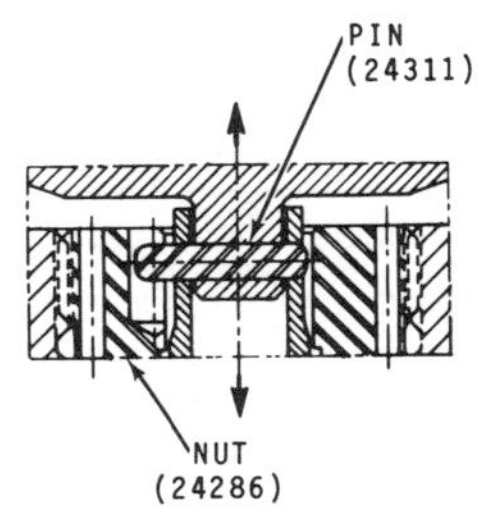	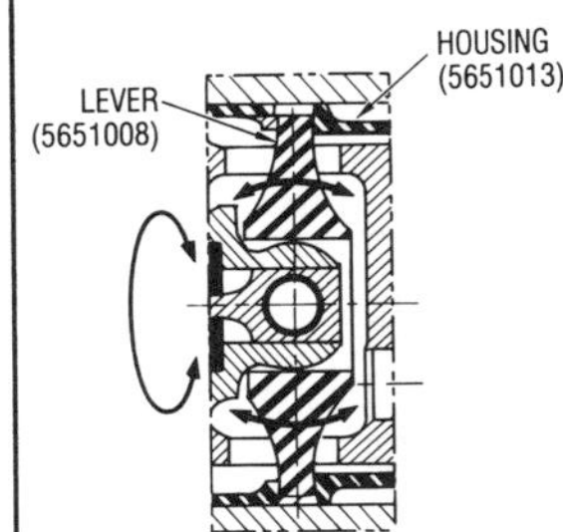	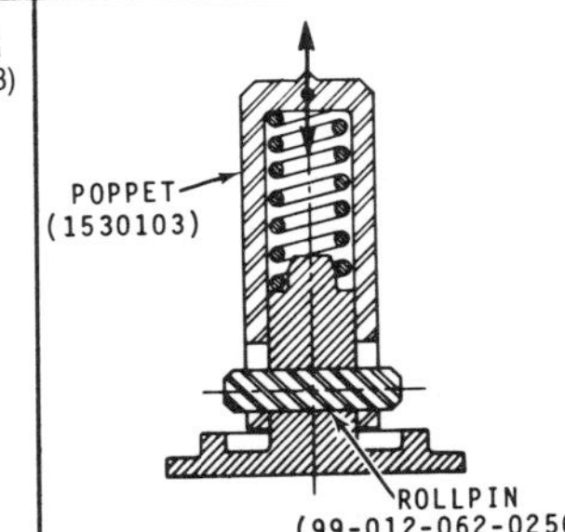	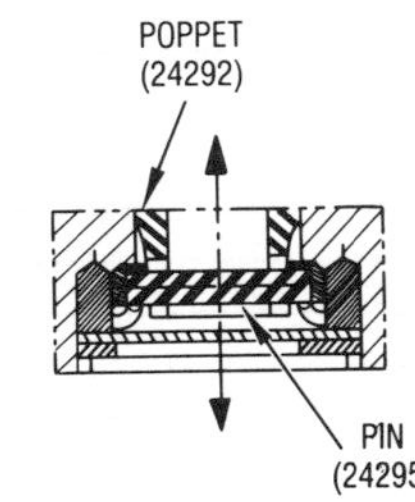
	GUIDE OR BEARING TYPE	POPPET ANTI – ROTATIONAL GUIDE	POPPET ANTI – ROTATIONAL GUIDE	POPPET ANTI – ROTATIONAL GUIDE	POPPET ANTI – ROTATIONAL GUIDE
	MATING MEMBERS CONTACT	LINE CONTACT	FACE CONTACT	LINE CONTACT	LINE CONTACT
	DESIGN OPERATING MEDIA	EXTERNAL ENVIRONMENT	GHe, GN_2 or AIR	GHe	EXTERNAL ENVIRONMENT
	DESIGN OPERATING TEMP. RANGE, °F	– 180 to + 160; – 200 to + 165	– 400 to + 165	– 180 to + 160	– 180 to + 160; – 200 to + 165
	DESIGN OPERATING LIFE CYCLES	2000	500	150	2000
	SPECIMEN OPERATION, CYCLES	1096; 5248	500	195	1096; 5248
INTERFACE MEMBERS DESCRIPTION — MOVABLE MEMBER	PART NUMBER	24311	5651008	1530103	24292
	PART NAME	PIN, POPPET	LEVER, BALL SCREW	POPPET – POPPET ASSY., R. V., REG.	POPPET, LOWER
	MATERIAL	MIL – S – 7720 303 CRES	AMS5737 A – 286 CRES	AMS5737 A – 286 CRES	QQ – S – 763 303 CRES
	HEAT TREATMENT	COND. A	R_C 28 MIN.	COND. H. T.	COND. A
	SURFACE FINISH	125 r m s	125 r m s	125 r m s	125 r m s
	SURFACE TREATMENT	PASSIVATE	PASSIVATE	DRY FILM LUBRICANT	ELECTROPOLISH
	OTHER SURFACE TREATMENTS	NONE	NONE	PASSIVATE	NONE
INTERFACE MEMBERS DESCRIPTION — STATIONARY MEMBER	PART NUMBER	24286	5651013	99 – 012 – 062 – 0250	24295
	PART NAME	NUT, LOCK	HOUSING – BALL SCREW	ROLLPIN	PIN, LOCKING
	MATERIAL	QQ – A – 325 6061 – T651 AL. ALY.	QQ – S – 763 303 CRES	ASTM B194 – 51 BeCu	QQ – W – 423 302 CRES
	HEAT TREATMENT	COND. – T651	AS REC'D. MAT'L. COND.	R_C 36 – 42	COND. B
	SURFACE FINISH	125 r m s	125 r m s	NOT SPECIFIED	AS REC'D. WIRE
	SURFACE TREATMENT	MIL – C – 5541 ALODINE # 1200	PASSIVATE	NONE	PASSIVATE
	OTHER SURFACE TREATMENTS	NONE	NONE	NONE	NONE
USING COMPONENT DESCRIPTION	PROCUREMENT DOCUMENT(S) OR P/N(S)	ME284 – 0356 – 0001; 60B52101 – 15F	20M32015	ME284 – 0158 – 0004	ME284 – 0356 – 0001; 60B52101 – 15F
	COMPONENT NAME	3 / 8 – INCH NORMALLY CLOSED SOLENOID VALVE	1 – 1 / 2 INCH HELIUM PRESSURE REGULATOR	3 / 8 – INCH HELIUM INJECTION REGULATOR – RELIEF VALVE	3 / 8 – INCH NORMALLY CLOSED SOLENOID VALVE
	MANUFACTURER	J. C. CARTER	PARKER – HANNIFIN	ROYAL INDUSTRIES	J. C. CARTER
	MANUFACTURER'S PART NUMBER	6969 – 2; 6969 – 15	5640061	12078	6969 – 2; 6969 – 15
	USING VEHICLE(S) AND STAGE(S)	SATURN S – II; SATURN S – IC	SATURN S – IC (QUALIFIED BACK–UP)	SATURN S – II	SATURN S – II; SATURN S – IC
	USING SYSTEM(S)	VALVE ACTUATION, HELIUM INJECTION, & ACCUMULATOR FILL	DNA	VALVE ACTUATION, HELIUM INJECTION, & ACCUMULATOR FILL	VALVE ACTUATION, HELIUM INJECTION, & ACCUMULATOR FILL

Interface Category Number 13
(Sheet 2 of 3)

MISCELLANEOUS LOCATING/SUPPORT GUIDES/BEARINGS

		DESIGN CONCEPT NUMBER	13-5	13-6	13-7	13-8
COMPONENT DATA BOOK		CONTROL NUMBER(S)	S – IVB / 209 – 2	S – II / 96 – 5	S – II / 96 – 5	S – IVB / 209 – 2
		INTERFACE DASH NUMBER(S)	– 4 & – 5	– 2 & – 4	– 1 & –3	– 7
		WEAR ANALYSIS RATING (AVG.)	**8.9**	**9.2**	**9.3**	**9.6**
		PHOTOGRAPHS, PAGES	B – 4, – 12, & – 13 '	B – 2, – 5, & –8	B – 2, – 5, & – 8	B – 5, & –14
INTERFACE DESCRIPTION		INTERFACE PICTORIAL DESCRIPTION	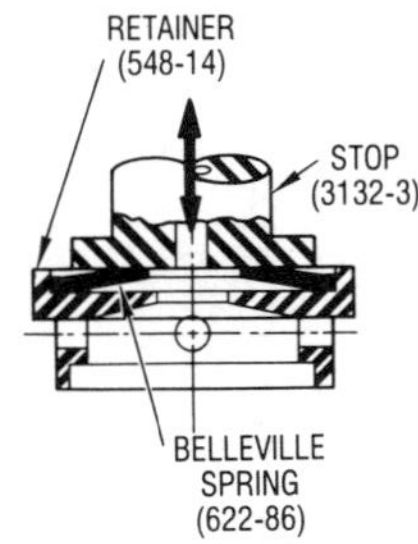	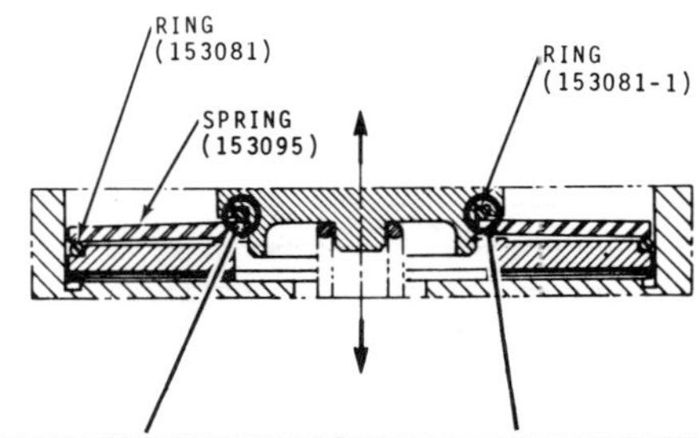		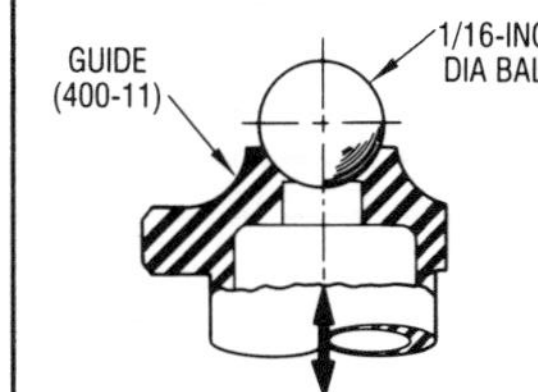
		GUIDE OR BEARING TYPE	BELLEVILLE SPRING SUPPORT	BELLEVILLE SPRING SUPPORT	BELLEVILLE SPRING SUPPORT	BALL GUIDE
		MATING MEMBERS CONTACT	LINE CONTACT (DIAMETRAL)	LINE CONTACT (DIAMETRAL)	LINE CONTACT (DIAMETRAL)	LINE CONTACT (DIAMETRAL)
		DESIGN OPERATING MEDIA	GHe	EXTERNAL ENVIRONMENT	EXTERNAL ENVIRONMENT	GHe
		DESIGN OPERATING TEMP. RANGE, °F	– 423 to + 160	– 180 to + 160	– 180 to + 160	– 423 to + 160
		DESIGN OPERATING LIFE CYCLES	5000	150; 500	150; 500	5000
		SPECIMEN OPERATION, CYCLES	10,121	195; 565	195/; 565	10,121
INTERFACE MEMBERS DESCRIPTION	MOVABLE MEMBER	PART NUMBER	622 – 86	153095	153095	NONE
		PART NAME	BELLEVILLE SPRING, SOL. VOLVE	SPRING – SENSOR, REG. R. V.	SPRING – SENSOR, REG. R. V.	1 / 16 DIA. BALL
		MATERIAL	MIL – S – 5059 301 CRES	NI – SPAN C 45 – 65% REDUCTION	NI – SPAN C 45 – 65% REDUCTION	440 C CRES
		HEAT TREATMENT	AS REC'D. MAT'L. COND.	TO ZERO THERMOELASTIC COEFF.	TO ZERO THERMOELASTIC COEFF.	UNKNOWN
		SURFACE FINISH	125 r m s	32 r m s	32 r m s	UNKNOWN
		SURFACE TREATMENT	PASSIVATE	NONE	NONE	NONE
		OTHER SURFACE TREATMENTS	NONE	NONE	NONE	NONE
	STATIONARY MEMBER	PART NUMBER(S)	3132 – 3; 548 – 14	153081 – 1	153081	400 – 11
		PART NAME(S)	STOP; RETAINER	RING – PIVOT, R. V., REG.	RING – PIVOT, R. V., REG	GUIDE, PILOT POPPET S. V.
		MATERIAL	18 – 8 CRES; 18 – 8 / 303 CRES	QQ – W – 423 18 – 8, 302 CRES	QQ – W – 423 18 – 8, 302 CRES	MIL – S – 7720 18 – 8 / 303 CRES
		HEAT TREATMENT	AS REC'D; COND. B	COND. B	COND. B	AS REC'D. MAT'L. COND.
		SURFACE FINISH	125 r m s	MILL FINISH	MILL FINISH	32 r m s
		SURFACE TREATMENT	PASSIVATE	DRY FILM LUBRICANT	DRY FILM LUBRICANT	PASSIVATE
		OTHER SURFACE TREATMENTS	NONE	NONE	NONE	NONE
USING COMPONENT DESCRIPTION		PROCUREMENT DOCUMENT(S) OR P/N(S)	1B43660 – 515	ME284 – 0158 – 0004	ME284 – 0158 – 0004	1B43660 – 515
		COMPONENT NAME	1 / 2 – INCH NORMALLY CLOSED HELIUM SOLENOID VALVE	3 / 8 – INCH HELIUM INJECTION REGULATOR – RELIEF VALVE	3 / 8 – INCH HELIUM INJECTION REGULATOR – RELIEF VALVE	1 / 2 – INCH NORMALLY CLOSED HELIUM SOLENOID VALVE
		MANUFACTURER	AMETEK / CALMEC	ROYAL INDUSTRIES	ROYAL INDUSTRIES	AMETEK / CALMEC
		MANUFACTURER'S PART NUMBER	548 – 515	12078	12078	548 – 515
		USING VEHICLE(S) AND STAGE(S)	SATURN S – IVB	SATURN S – II	SATURN S – II	SATURN S – IVB
		USING SYSTEM(S)	PRESSURE CONTROL	VALVE ACTUATION, HELIUM INJECTION, & ACCUMULATOR FILL	VALVE ACTUATION, HELIUM INJECTION, & ACCUMULATOR FILL	PRESSURE CONTROL

Interface Category Number 13 (Sheet 3 of 3)

MISCELLANEOUS LOCATING/SUPPORT GUIDES/BEARINGS

DESIGN CONCEPT NUMBER		13-9			
COMPONENT DATA BOOK	CONTROL NUMBER(S)	S - II 90 - 1			
	INTERFACE DASH NUMBER(S)	- 7 & - 8			
	WEAR ANALYSIS RATING (AVG.)	**9.8**			
	PHOTOGRAPHS, PAGES	B - 5, - 6, & - 7			
INTERFACE DESCRIPTION	INTERFACE PICTORIAL DESCRIPTION	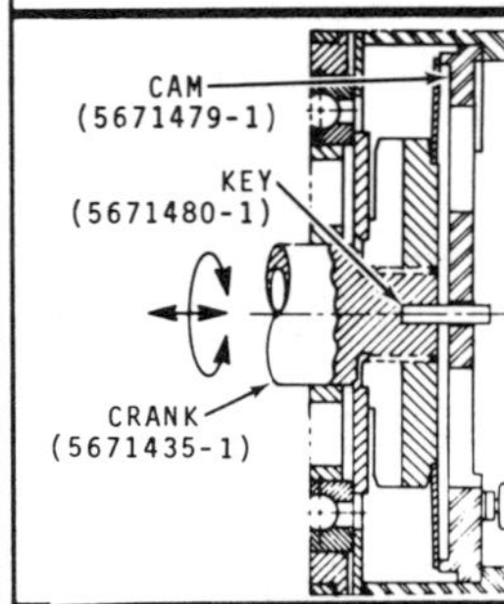			
	GUIDE OR BEARING TYPE	KEYWAY FACES			
	MATING MEMBERS CONTACT	FACE CONTACT			
	DESIGN OPERATING MEDIA	LOX or LH_2			
	DESIGN OPERATING TEMP. RANGE, °F	- 423 to + 160			
	DESIGN OPERATING LIFE CYCLES	1000			
	SPECIMEN OPERATION, CYCLES	1178			
INTERFACE MEMBERS DESCRIPTION — MOVABLE MEMBER	PART NUMBER(S)	5671435 - 1; 5671479 - 1			
	PART NAME(S)	CRANK; CAM			
	MATERIAL	15 - 5PH CRES; 7075 - T6 AL. ALY.			
	HEAT TREATMENT	R_c 24 / 30; COND. - T6			
	SURFACE FINISH	125 r m s			
	SURFACE TREATMENT	PASSIVATE; HARD ANODIZE			
	OTHER SURFACE TREATMENTS	NONE; OVERSPRAY LUBECO # 905			
INTERFACE MEMBERS DESCRIPTION — STATIONARY MEMBER	PART NUMBER	5671480 - 1			
	PART NAME	KEY			
	MATERIAL	MIL - S - 5059 301, 302, or 316 CRES			
	HEAT TREATMENT	1 / 2 HARD MIN.			
	SURFACE FINISH	125 r m s			
	SURFACE TREATMENT	PASSIVATE			
	OTHER SURFACE TREATMENTS	NONE			
USING COMPONENT DESCRIPTION	PROCUREMENT DOCUMENT(S) OR P/N(S)	ME284 - 0358 - 0001			
	COMPONENT NAME	8 - INCH PROPELLANT SHUTOFF VALVE (PREVALVE)			
	MANUFACTURER	PARKER - HANNIFIN			
	MANUFACTURER'S PART NUMBER	5640024			
	USING VEHICLE(S) AND STAGE(S)	SATURN S - II			
	USING SYSTEM(S)	LOX & LH_2 FEED			

Interface Category Number 14 (Sheet 1 of 4)

AXIAL STOPS AND IMPACT MEMBERS (PUSH-RODS)

	DESIGN CONCEPT NUMBER	14-1	14-2	14-3	14-4
COMPONENT DATA BOOK	CONTROL NUMBER(S)	S-IVB/210-3	S-II/100-1	S-II/98-5	S-II/98-5
	INTERFACE DASH NUMBER(S)	-9 & -9A	-10	-6	-7
	WEAR ANALYSIS RATING (AVG.)	**6.4**	**6.7**	**6.9**	**7.6**
	PHOTOGRAPHS, PAGES	B-3, -4, -5, & -13	82, 92, & 93	B-2, & -13	B-2, & -13
INTERFACE DESCRIPTION	INTERFACE PICTORIAL DESCRIPTION	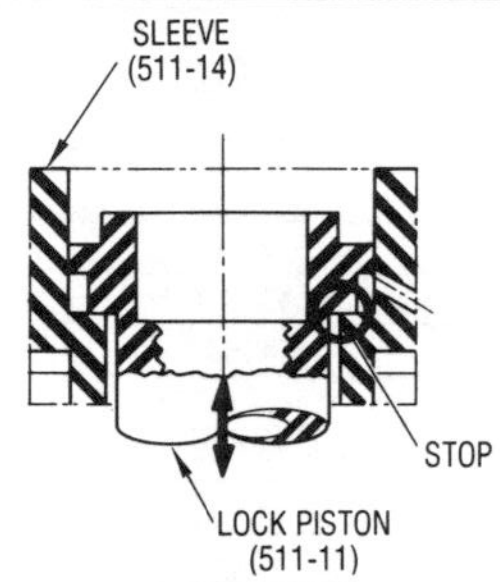	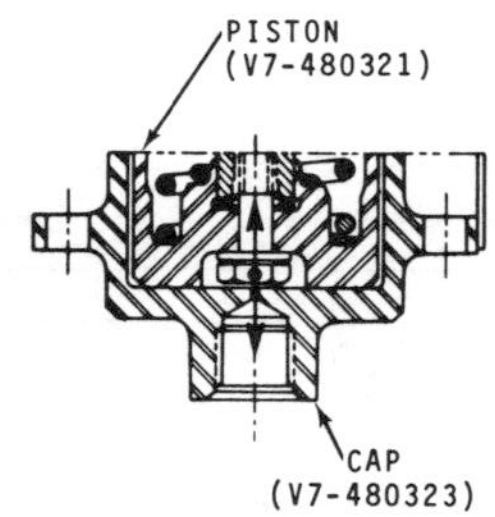	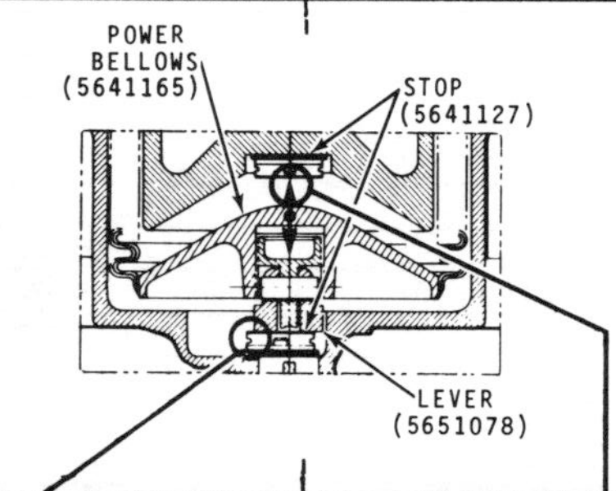	
	BASIC INTERFACE FUNCTION	LOCK PISTON STOP	ACTUATOR PISTON STOP	DRIVE LINKAGE STOP	ACT. POWER BELLOWS STOP
	STOP OR IMPACT MEMBER CONTACT	DIAMETRAL CONTACT RING	DIAMETRAL CONTACT RING	FLAT FACE CONTACT	POINT CONTACT (SPHERICAL)
	DESIGN OPERATING MEDIA	GHe	GHe	GOX	GOX
	DESIGN OPERATING TEMP. RANGE, °F	-423 to +160	-423 to +160	-180 to +860	-180 to +860
	DESIGN OPERATING LIFE, CYCLES	200	750	2000	2000
	SPECIMEN OPERATION, CYCLES	30	935	147	147
INTERFACE MEMBERS DESCRIPTION — MOVABLE MEMBER	PART NUMBER	511-11	V7-480321	5651078	5641165
	PART NAME	PISTON-LOCK	PISTON, RECIRC. VALVE, N. O.	LEVER - STABILIZING	POWER BELLOWS
	MATERIAL	AMS 5643 17-4 PH CRES	AMS 5737 A-286 CRES	AMS 5737 A-286 CRES	AMS 5645 321 CRES
	HEAT TREATMENT	COND. H-925	AS REC'D. MAT'L. COND.	AS REC'D. MAT'L. COND.	AS REC'D. MAT'L. COND.
	SURFACE FINISH	125 r m s	125 r m s	125 r m s	125 r m s
	SURFACE TREATMENT	PASSIVATE	PASSIVATE	PASSIVATE	PASSIVATE
	OTHER SURFACE TREATMENTS	NONE	NONE	NONE	NONE
INTERFACE MEMBERS DESCRIPTION — STATIONARY MEMBER(S)	PART NUMBER	511-14	V7-480323	5641127	5641127
	PART NAME	SLEEVE, BALL RETAINER	CAP, ACT. - RECIRC. VALVE, N. O.	STOP, BELLOWS	STOP, BELLOWS
	MATERIAL	AMS 5643 17-4PH CRES	QQ-A-225/8 6061-T651 AL. ALY.	MIL-S-7720 303 Se CRES	MIL-S-7720-303 Se CRES
	HEAT TREATMENT	COND. H-925	COND. -T651	COND. B	COND. B
	SURFACE FINISH	125 r m s	125 r m s	125 r m s	125 r m s
	SURFACE TREATMENT	PASSIVATE	CHROMIC ACID ANODIZE	PASSIVATE	PASSIVATE
	OTHER SURFACE TREATMENTS	NONE	NONE	NONE	NONE
USING COMPONENT DESCRIPTION	PROCUREMENT DOCUMENT(S) OR P/N(S)	1A49988	V7-480328-91	ME284-0161-0008	ME248-0161-0008
	COMPONENT NAME	6-INCH DIRECTIONAL CONTROL FUEL VENT VALVE	2-INCH NORMALLY OPEN RECIRCULATION VALVE	3-INCH GOX PRESSURE REGULATOR	3-INCH GOX PRESSURE REGULATOR
	MANUFACTURER	AMETEK/CALMEC	ROCKWELL/LAD	PARKER-HANNIFIN	PARKER/HANNIFIN
	MANUFACTURER'S PART NUMBER	511-509	V7-480328-91	5640012M7	5640012M7
	USING VEHICLE(S) AND STAGE(S)	SATURN-S-IVB	SATURN S-II	SATURN S-II	SATURN S-II
	USING SYSTEM(S)	LH_2 TANK VENT	LOX & LH_2 RECIRCULATION	PREVIOUSLY IN LOX TANK PRESSURIZATION	PREVIOUSLY IN LOX TANK PRESSURIZATION

Interface Category Number 14
(Sheet 2 of 4)

AXIAL STOPS AND IMPACT MEMBERS (PUSH-RODS)

DESIGN CONCEPT NUMBER		14-5	14-6	14-7	14-8
COMPONENT DATA BOOK	CONTROL NUMBER(S)	S – IVB / 209 – 2	S – II / 89 – 1	S – II / 98 – 5	S – IVB / 210 – 3
	INTERFACE DASH NUMBER(S)	– 1	– 14 & – 18	– 3	– 11 & – 11A
	WEAR ANALYSIS RATING (AVG.)	**8.0**	**8.7**	**8.7**	**9.1**
	PHOTOGRAPHS, PAGES	B – 4, & – 7	B – 5, – 6, – 15, & – 20	B – 5 & – 11	B – 3, – 4, – 6, & – 15
INTERFACE DESCRIPTION	INTERFACE PICTORIAL DESCRIPTION	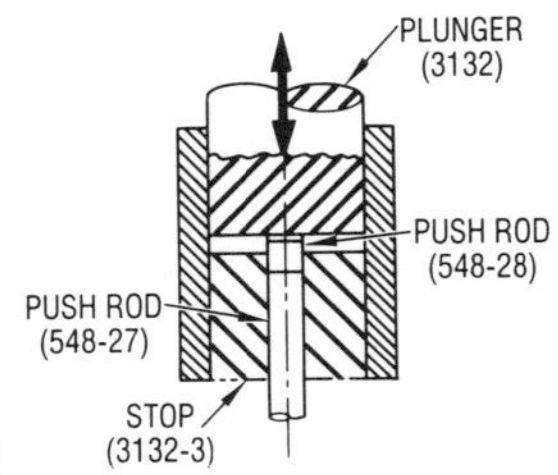	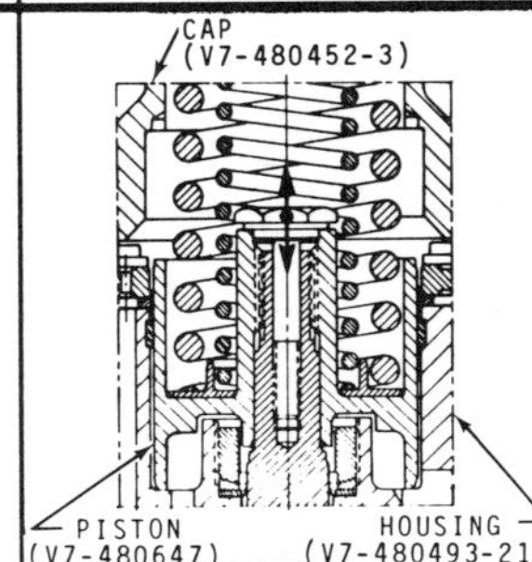	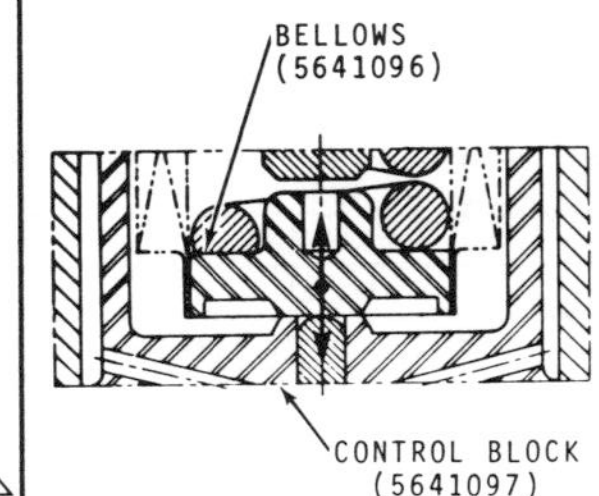	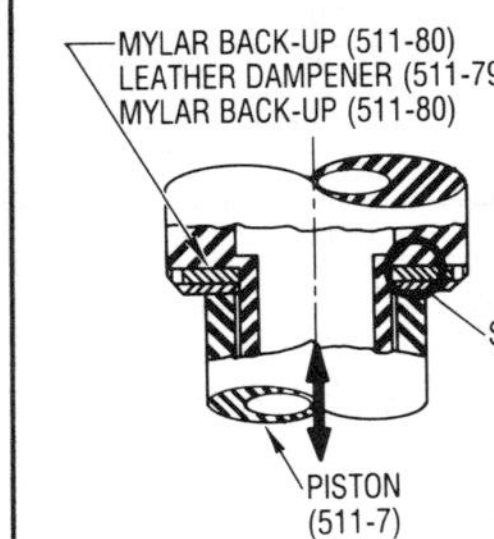
	BASIC INTERFACE FUNCTION	PILOT SEAT PUSH – ROD	ACTUATOR PISTON STOPS	ACTUATOR BELLOWS STOP	ACTUATOR PISTON STOPS
	STOP OR IMPACT MEMBER CONTACT	FLAT FACE CONTACT	DIAMETRAL CONTACT RINGS	DIAMETRAL CONTACT RING	DIAMETRAL CONTACT RINGS
	DESIGN OPERATING MEDIA	GHe	GHe	GOX	GHe
	DESIGN OPERATING TEMP. RANGE, °F	– 423 to + 160	– 423 to + 160	– 180 to + 860	– 423 to + 160
	DESIGN OPERATING LIFE, CYCLES	5000	750	2000	200
	SPECIMEN OPERATION, CYCLES	10,121	591	147	30
INTERFACE MEMBERS DESCRIPTION – MOVABLE MEMBER	PART NUMBER	3132	V7 – 480647	5641096	511 – 7
	PART NAME	PLUNGER, SOLENOID ASSY.	PISTON – ACT., FILL VALVE	BELLOWS – BIAS REGULATOR	PISTON – CONTROL
	MATERIAL	1213C LOW CARBON STEEL	MB0160 – 020 304L CRES	AMS 5646 347 CRES	AMS 5643 17 – 4 PH CRES
	HEAT TREATMENT	NOT SPECIFIED	AS REC'D. MAT'L. COND.	AS REC'D. MAT'L. COND.	COND. H – 925
	SURFACE FINISH	125 r m s	125 r m s	125 r m s	32 r m s
	SURFACE TREATMENT	NICKEL PLATED	PASSIVATE	PASSIVATE	PASSIVATE
	OTHER SURFACE TREATMENTS	NONE	NONE	NONE	NONE
INTERFACE MEMBERS DESCRIPTION – STATIONARY MEMBER(S)	PART NUMBER(S)	548 – 28	V7 – 480493 – 21; V7 – 480452 – 3	5641097	511 – 80; 511 – 79
	PART NAME(S)	ROD, SOLENOID ARMATURE	HOUSING; CAP	CONTROL BLOCK – BIAS REG.	BACK – UP; DAMPENER
	MATERIAL	MIL – S – 7720 18 – 8 / 303 CRES	TENS 50 – T60 AL ALY; 6061 – T651 AL ALY	MIL – S – 7720 303 Se CRES	MYLAR A or D; MIL – R – 5521 LEATHER
	HEAT TREATMENT	COND. B	COND. – T60; COND. – T651	COND. B	AS REC'D. MAT'L. COND.
	SURFACE FINISH	125 r m s	125 r m s	125 r m s	125 r m s
	SURFACE TREATMENT	PASSIVATE	ANODIZE	PASSIVATE	NONE
	OTHER SURFACE TREATMENTS	NONE	NONE	NONE	NONE
USING COMPONENT DESCRIPTION	PROCUREMENT DOCUMENT(S) OR P/N(S)	1B43660 – 515	V7 – 480450 – 121	ME284 – 0161 – 0008	1A49988
	COMPONENT NAME	1 / 2 – INCH NORMALLY CLOSED SOLENOID VALVE	8 – INCH PROPELLANT FILL & DRAIN VALVE	3 – INCH GOX PRESSURE REGULATOR	6 – INCH DIRECTIONAL CONTROL FUEL VENT VALVE
	MANUFACTURER	AMETEK / CALMEC	ROCKWELL / LAD	PARKER – HANNIFIN	AMETEK / CALMEC
	MANUFACTURER'S PART NUMBER	548 – 515	V7 – 480450 – 121	5640012M7	511 – 509
	USING VEHICLE(S) AND STAGE(S)	SATURN S – IVB	SATURN S – II	SATURN S – II	SATURN S – IVB
	USING SYSTEM(S)	PRESSURE CONTROL	LOX & LH_2 TANK FILL & DRAIN	PREVIOUSLY IN LOX TANK PRESSURIZATION	LH_2 TANK VENT

Interface Category Number 14
(Sheet 3 of 4)

AXIAL STOPS AND IMPACT MEMBERS (PUSH-RODS)

		DESIGN CONCEPT NUMBER	14-9	14-10	14-11	14-12
COMPONENT DATA BOOK		CONTROL NUMBER(S)	S – II / 98 – 5	S – II / 96 – 5	S – IC / 206 – 1	J – 2 / 208 – 1
		INTERFACE DASH NUMBER(S)	– 19	– 5 & – 13	– 18 & – 19	– 14 & – 16
		WEAR ANALYSIS RATING (AVG.)	**9.2**	**9.3**	**9.4**	**9.5**
		PHOTOGRAPHS, PAGES	B – 8, & – 22	B – 2, – 5,, & – 9	B – 4, & – 15	B – 5, – 6, – 7, – 18, & – 20
INTERFACE DESCRIPTION		INTERFACE PICTORIAL DESCRIPTION	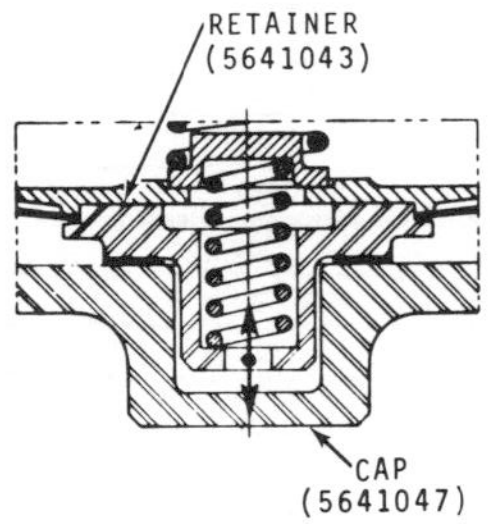	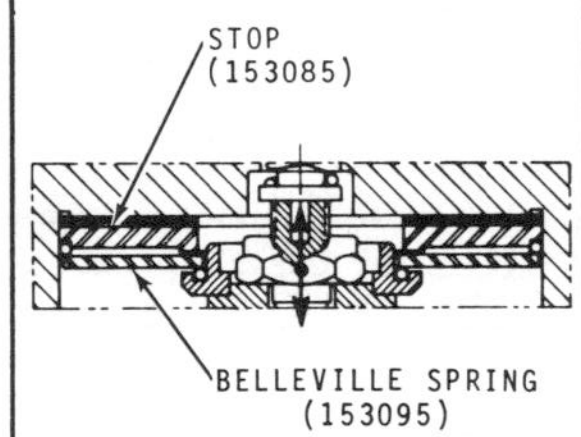	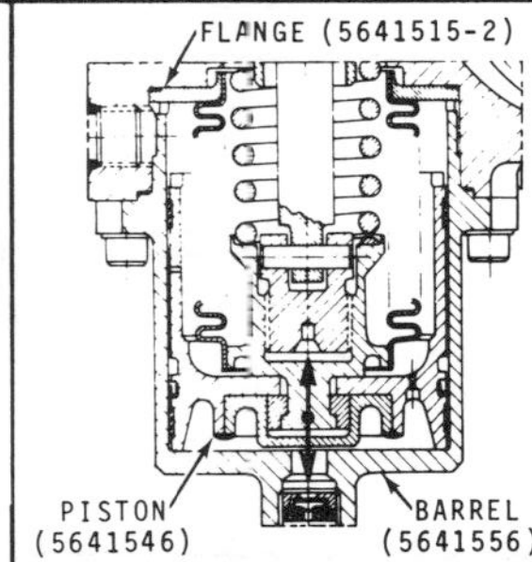	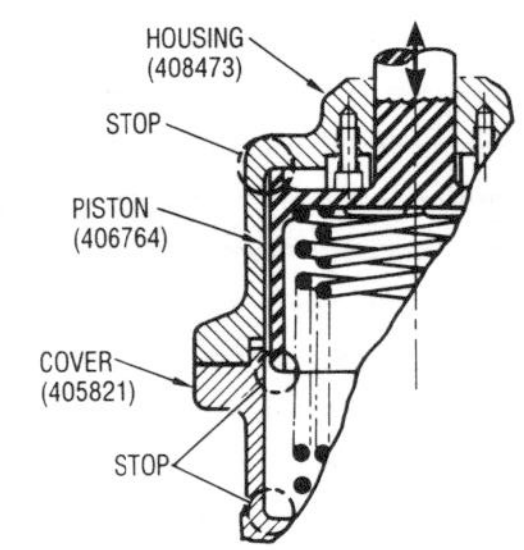
		BASIC INTERFACE FUNCTION	SOLENOID VALVE SEAT STOP	BELLEVILLI SPRING STOPS	ACTUATOR PISTON STOPS	ACTUATOR PISTON STOPS
		STOP OR IMPACT MEMBER CONTACT	DIAMETRAL CONTACT RING	DIAMETRAL CONTACT RINGS	DIAMETRAL CONTACT RINGS	DIAMETRAL CONTACT RINGS
		DESIGN OPERATING MEDIA	GOX	EXTERNAL ENVIRONMENT	LH_2	GHe
		DESIGN OPERATING TEMP. RANGE, °F	– 180 to + 860	– 180 to + 160	– 423 to + 165	– 420 to + 140
		DESIGN OPERATING LIFE, CYCLES	2000	150; 500	2000	1308
		SPECIMEN OPERATION, CYCLES	147	195; 565	3318	292
INTERFACE MEMBERS DESCRIPTION	MOVABLE MEMBER(S)	PART NUMBER	5641043	153095	5641546	406764
		PART NAME	RETAINER – SPRING, PLUNGER	SPRING – SENSOR, REG., R. V.	PISTON	PISTON, MPV
		MATERIAL	MIL – S – 7720 316 CRES	NI – SPAN C 45 – 65 % REDUCTION	QQ – A – 225 / 8 6061 – T651 AL. ALY.	A – 286 CRES
		HEAT TREATMENT	COND. A	TO ZERO THERMOPLASTIC COEFF.	COND. – T651	AS REC'D. MAT'L. COND.
		SURFACE FINISH	125 r m s	32 r m s	125 r m s	32 r m s
		SURFACE TREATMENT	PASSIVATE	NONE	MIL – A – 8625 ANODIZE	NONE
		OTHER SURFACE TREATMENTS	NONE	NONE	NONE	NONE
	STATIONARY MEMBER(S)	PART NUMBER(S)	5641047	153085	5641556; 5641515 – 2	408473; 405821
		PART NAME(S)	CAP	STOP – BELLEVILLE, R. V., REG.	BARREL; FLANGE	HOUSING; COVER
		MATERIAL	MIL – S – 7720 316 CRES	QQ – S – 763 303 S / Se CRES	6061 – T651 AL. ALY.; 321 CRES	(ROUGH CASTING); TENS 50–T6 AL ALY
		HEAT TREATMENT	COND. A	COND. A / B	COND. – T651; AS REC'D. MAT'L.	–T6 COND.; –T6 COND.
		SURFACE FINISH	125 r m s	125 r m s	125 r m s	125 r m s
		SURFACE TREATMENT	PASSIVATE	PASSIVATE	ANODIZE; PASSIVATE	HARD ANODIZE; ANODIZE
		OTHER SURFACE TREATMENTS	NONE	NONE	NONE	NONE
USING COMPONENT DESCRIPTION		PROCUREMENT DOCUMENT(S) OR P/N(S)	ME284 – 0161 – 0008	ME284 – 0158 – 0004	20M32014	409920X7
		COMPONENT NAME	3 – INCH GOX PRESSURE REGULATOR	3 / 8 – INCH HELIUM INJECTION REGULATOR – RELIEF VALVE	2 – INCH NORMALLY CLOSED SUBMERGED LH_2 SHUTOFF VALVE	4 – INCH NORMALLY CLOSED MAIN FUEL VALVE
		MANUFACTURER	PARKER – HANNIFIN	ROYAL INDUSTRIES	PARKER – HANNIFIN	ROCKWELL / RKD
		MANUFACTURER'S PART NUMBER	5640012M7	12078	5640027	409920X7
		USING VEHICLE(S) AND STAGE(S)	SATURN S – II	SATURN S – II	SATURN S – IC (QUALIFIED BACKUP)	SATURN S – II & SATURN S – IVB
		USING SYSTEM(S)	PREVIOUSLY IN LOX TANK PRESSURIZATION	VALVE ACTUATOR, HELIUM INJECTION, & ACCUMULATOR FILL	DNA	J – 2 ENGINE LH_2 FEEL

Interface Category Number 14
(Sheet 4 of 4)

AXIAL STOPS AND IMPACT MEMBERS (PUSH-RODS)

		DESIGN CONCEPT NUMBER	14-13	14-14		
COMPONENT DATA BOOK		CONTROL NUMBER(S)	J-2/208-1	S-IVB/209-2		
		INTERFACE DASH NUMBER(S)	-19	-15		
		WEAR ANALYSIS RATING (AVG.)	**9.8**	**10.0**		
		PHOTOGRAPHS, PAGES	B-6, -7, -20, & -23	B-4, -5, & -19		
INTERFACE DESCRIPTION		INTERFACE PICTORIAL DESCRIPTION	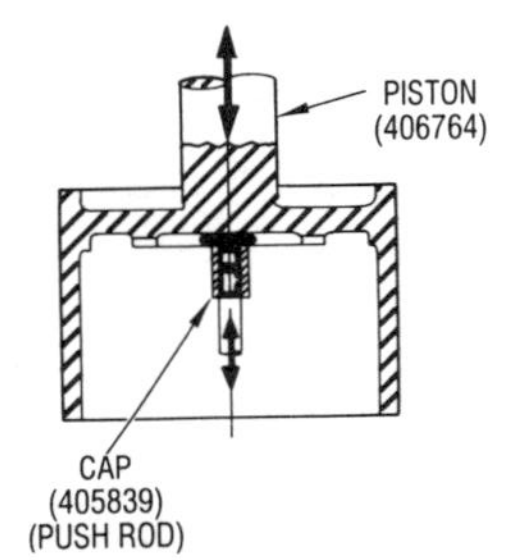	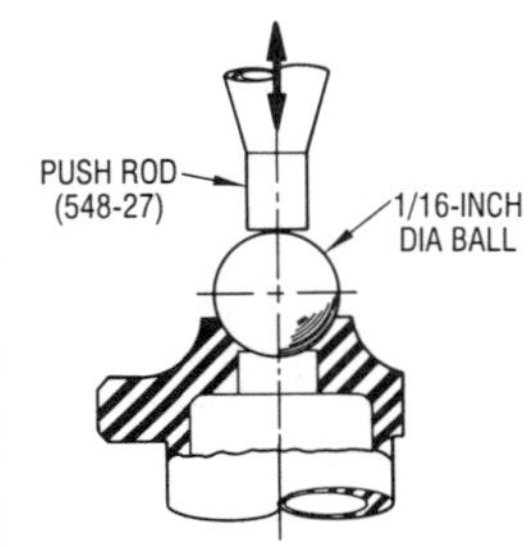		
		BASIC INTERFACE FUNCTION	SEQUENCE VALVE PUSH-ROD	PILOT VALVE PUSH-ROD		
		STOP OR IMPACT MEMBER CONTACT	FLAT FACE CONTACT	POINT (SPHERICAL)		
		DESIGN OPERATING MEDIA	GHe	GHe		
		DESIGN OPERATING TEMP. RANGE, °F	-420 to +140	-423 to +160		
		DESIGN OPERATING LIFE, CYCLES	1308	5000		
		SPECIMEN OPERATION, CYCLES	292	10,121		
INTERFACE MEMBERS DESCRIPTION	MOVABLE MEMBER	PART NUMBER	406764	548-27		
		PART NAME	PISTON, MPV	ROD, SOLENOID PUSH		
		MATERIAL	A-286 CRES	MIL-S-7720 18-8/303 CRES		
		HEAT TREATMENT	AS REC'D. MAT'L. COND.	COND. B		
		SURFACE FINISH	125 r m s	125 r m s		
		SURFACE TREATMENT	NONE	PASSIVATE		
		OTHER SURFACE TREATMENTS	NONE	NONE		
	STATIONARY MEMBER(S)	PART NUMBER	405839	NONE		
		PART NAME	CAP, SEQUENCE VALVE	1/16 DIA. BALL		
		MATERIAL	AMS 5737 A-286 CRES	440 C CRES		
		HEAT TREATMENT	R_c 29-40	NOT SPECIFIED		
		SURFACE FINISH	63 r m s	NOT SPECIFIED		
		SURFACE TREATMENT	NONE	NOT SPECIFIED		
		OTHER SURFACE TREATMENTS	NONE	NONE		
USING COMPONENT DESCRIPTION		PROCUREMENT DOCUMENT(S) OR P/N(S)	409920X7	1B43660-515		
		COMPONENT NAME	4-INCH NORMALLY CLOSED MAIN FUEL VALVE	1/2-INCH NORMALLY CLOSED SOLENOID VALVE		
		MANUFACTURER	ROCKWELL/RKD	AMETEK/CALMEC		
		MANUFACTURER'S PART NUMBER	409920X7	548-515		
		USING VEHICLE(S) AND STAGE(S)	SATURN S-II & SATURN S-IVB	SATURN S-IVB		
		USING SYSTEM(S)	J-2 ENGINE LH_2 FEED	PRESSURE CONTROL		

Interface Category Number 15
(Sheet 1 of 1)

BALL LOCKS (DETENTS)

DESIGN CONCEPT NUMBER			15-1			
COMPONENT DATA BOOK		CONTROL NUMBER(S)	S-IVB/210-3			
		INTERFACE DASH NUMBER(S)	-6 & -6A			
		WEAR ANALYSIS RATING (AVG.)	**7.1**			
		PHOTOGRAPHS, PAGES	B-3, -4, -5, -10, & -11			
INTERFACE DESCRIPTION		INTERFACE PICTORIAL DESCRIPTION	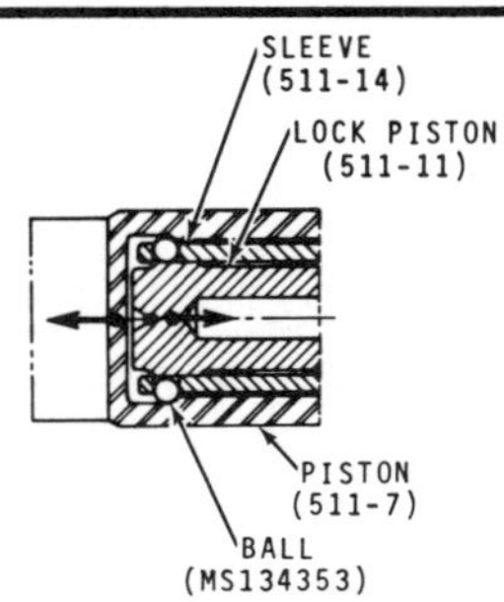			
		DETENT TYPE	BALL LOCK			
		LOCKED MEMBER	ACTUATOR PISTON			
		DESIGN OPERATING MEDIA	GHe			
		DESIGN OPERATING TEMP. RANGE, °F	-423 to +160			
		DESIGN OPERATING LIFE, CYCLES	200			
		SPECIMEN OPERATION, CYCLES	30			
INTERFACE MEMBERS DESCRIPTION	BALL	PART NUMBER	MS134353			
		PART NAME	BALL LOCK			
		MATERIAL	AMS7445 440C CRES			
		HEAT TREATMENT	Rc 55-64			
		SURFACE FINISH	2 r m s			
		SURFACE TREATMENT	NOT SPECIFIED			
		OTHER SURFACE TREATMENTS	NONE			
	MATING MEMBER(S)	PART NUMBER(S)	511-11; 511-7; 511-14			
		PART NAME(S)	PISTON; PISTON; SLEEVE			
		MATERIAL	AMS5643 17-4PH CRES			
		HEAT TREATMENT	COND. H-925			
		SURFACE FINISH	32 r m s			
		SURFACE TREATMENT	PASSIVATE			
		OTHER SURFACE TREATMENTS	NOINE			
USING COMPONENT DESCRIPTION		PROCUREMENT DOCUMENT(S) OR P/N(S)	1A49988			
		COMPONENT NAME	8-INCH DIRECTIONAL CONTROL FUEL VENT VALVE			
		MANUFACTURER	AMETEK/CALMEC			
		MANUFACTURER'S PART NUMBER	511-509			
		USING VEHICLE(S) AND STAGE(S)	SATURN S-IVB			
		USING SYSTEM(S)	LH_2 TANK VENT			